D1526565

"Mike Buratovich has provided a helpful discussion on a variety of longstanding as well as emerging controversies on bioethics. He consults both ancient and modern voices from ethics, philosophy, and theology and lays them aside current discoveries and developments in biological science. He quotes those from both sides of the issues. The result is a splendid, comprehensive, and detailed treatment of sensitive matters of abortion, personhood, stem cell research, and cloning. He does it with fairness, sensitivity, and objectivity—those graces missing from much of public discourse. I trust believers and nonbelievers and pro-life and pro-choice proponents alike will benefit from this book. Each chapter is organized as a simulated teacher-student interchange around a specific controversy. The chapters provide a helpful catalog of issues that can be referenced on their own for what the author hopes will result in an increase in thoughtful, informed, and biblical voices in the public debate."

—Mark Van Valin,
Pastor, Spring Arbor Free Methodist Church

The Stem Cell Epistles

The Stem Cell Epistles

Letters to My Students about Bioethics, Embryos, Stem Cells, and Fertility Treatments

MICHAEL A. BURATOVICH

Foreword by Congressman Timothy Walberg

CASCADE *Books* • Eugene, Oregon

THE STEM CELL EPISTLES
Letters to My Students about Bioethics, Embryos,
Stem Cells, and Fertility Treatments

Cascade Books
An Imprint of Wipf and Stock Publishers
199 W. 8th Ave., Suite 3
Eugene, OR 97401

www.wipfandstock.com

ISBN 13: 978-1-62032-276-5

Cataloguing-in-Publication data:

Buratovich, Michael A.

The stem cell epistles : letters to my students about bioethics, embryos, stem cells, and fertility treatments / Michael A. Buratovich ; foreword by Congressman Timothy Walberg.

xxiv + 250 pp. ; 23 cm. Includes bibliographical references and index.

ISBN 13: 978-1-62032-276-5

1. Embryonic stem cells—Research—Moral and ethical aspects—United States. 2. Embryonic stem cells—Research—United States—Religious aspects. 3. Embryonic stem cells—Research—Political aspects—United States. I. Walberg, Timothy. II. Title.

QH588.S83 B87 2013

Manufactured in the U.S.A.

To my eldest daughter, Rachel,
whose love of the written word inspired me to write.

Contents

Illustrations and Tables

Illustrations

Tables

Foreword

As a member of the United States Congress, I am frequently asked to read this or that book, pamphlet, or Web site and render an opinion on it. The authors are looking for an endorsement, which is fine, but it makes enormous demands on the time of a legislator. Therefore, I have become somewhat cautious when people ask me to read their latest thoughts, regardless of the format in which they are presented.

With this in mind, in March 2010 I was offered a manuscript to read by a member of my constituency, which is Michigan's 7th Congressional District. The author was a gentleman I had met a few times at pro-life meetings and at one of the churches my wife and I regularly visit. He and I share a common conviction of the sanctity of human persons at every stage of life. However, the author of this manuscript, Dr. Michael Buratovich, came to this conviction in a slightly different way than I did. Michael, you see, is trained as a developmental biologist who teaches human embryology at the university level. When it comes to the unborn, he brings his scientific training to bear on the question of the status of the unborn.

With that in mind, Michael had penned a manuscript that consisted of questions his students, colleagues, and even some parents had asked him about the science, ethics, and potential application of embryonic and adult stem cells. Because the issue of embryonic stem cells recurrently confronts legislators, having an informed opinion on it is essential. Therefore, I was intrigued, though slightly skeptical. I originally trained and served as a Christian pastor for a decade before becoming a legislator. While I have had a few college courses in science, and have read rather broadly about it, I am not an expert on scientific issues and dreaded wading into a thick tome filled with mind-numbing details about stem cells. However, once I began reading the manuscript, I found myself drawn into the stem cell issue. The manuscript detailed each question and the author's careful, thoughtful, and well-documented answer. Here was a very readable response to some very good questions I had asked many times. All in all, reading it was a remarkable journey.

Since reading it, I have borrowed Michael's thoughts several times when the stem cell issue came up in my conversations with congressional colleagues, staffers, members of my constituency from Michigan, or friends. I am grateful for the opportunity to read Michael's manuscript because it is has proven its worth several times over. Now that it has finally been published in book form by Wipf and Stock Publishers, you too have the opportunity to experience the informative and enjoyable journey I experienced a few years ago. In this book is a solid contribution to the stem cell debate that no one, regardless of his or her opinion, should miss.

I am grateful that Michael saw the need to write this book about a pressing issue competing for the soul of our nation. Dig into this book and digest it. It will reward you several times over.

Timothy Walberg (R), Congressman
7th Congressional District, Michigan

Preface

IN 1981, TWO LABORATORIES independently made the same revolutionary discovery. Gail Martin at the University of California, San Francisco, and Martin Evans and Matthew Kaufman at Cambridge University isolated embryonic stem cells from early mouse embryos.[1] These cell lines transformed the study of developmental biology. Because they were isolated from early embryos, embryonic stem cells could be placed into an embryo and integrate into the developing mouse. Genetic manipulation of mouse embryonic stem cells could produce animals that were missing specific genes in particular tissues and even at various times during development, which gave vital clues to gene function. Genetic diseases were studied with more exacting precision, and scientists could create mice with genomes they had designed.[2]

In 1995, James A. Thomson at the Wisconsin Regional Primate Research Center at the University of Wisconsin, Madison, reported the derivation of embryonic stem cells from rhesus monkey embryos,[3] and in 1996, he published a paper that announced the production of similar cells from early embryos of the common marmoset.[4] The generation of embryonic stem cells from nonhuman primates was the prelude to deriving such cells from human embryos. Because the Dickey-Wicker Amendment prevented federal funding for research that destroyed human embryos, Thomson had to work in sections of that lab that were completely funded by private funding. Thomson's funding came from Geron Corporation in Menlo Park, California, and in November 1998, Thomson published a paper that announced the isolation of human embryonic stem cells.[5]

1. Martin, "Pluripotent Cell Line from Early Mouse Embryos," 7634–38; Evans and Kaufman, "Pluripotential Cells from Mouse Embryos," 154–56.

2. Cohen-Tannoudii and Babinet, "Beyond 'Knock-Out' Mice," 929–38.

3. Thomson et al., "Primate Embryonic Stem Cells," 7844–88.

4. Thomson et al., "Pluripotent Cell Lines from Common Marmoset," 254–59.

5. Thomson et al., "Embryonic Stem Cell Lines from Human Blastocysts," 1145–47.

From the beginning, the derivation of human embryonic stem cells was hailed as a great scientific breakthrough,[6] and it certainly was. However, it was also a troubling achievement. In order to make the embryonic stem cell lines, human early embryos had to be destroyed. If those same embryos were implanted into a mother's womb, they probably would have resulted in the birth of a baby that would grow into a toddler, an adolescent, a young adult, perhaps a parent and grandparent. The essential connection between adult human beings and the embryos that were being dismembered is obvious. Should we be destroying the youngest and most defenseless members of our communities?

Such a concern was often dismissed as quaint, unnecessary, Luddite, or even "religious fanaticism."[7] Nevertheless, the opposition to this research did not come solely from laypeople, but scientists were also divided on this issue. Some openly questioned, "Just because we can do this, does that mean we should?"[8]

I watched from the sidelines for some time, but it was not because I had nothing to say; rather, I was waiting for the right time to say it. I read and thought about this issue from the perspective of someone who has studied developmental biology at the graduate level. In my consideration of the issue, I became convinced that the embryo is the youngest member of the human race. While human embryos do not have all the rights an adult might have, it has at least the right not to be harmed.

Furthermore, despite all the focus on embryonic stem cells, a renaissance in medicine is quietly being missed as treatments for many severe maladies are being fashioned from stem cells from umbilical cord tissues, umbilical cord blood, and from adult human bodies. Many treatments with these new tools have been developed or are in the formative stages of development. These discoveries are changing the face of medicine and have ushered in a new field of modern medicine called regenerative medicine. Regenerative medicine includes using stem cells from the patient's own body or a donor to treat various maladies. It also includes the remarkable discipline known as tissue engineering, in which moldable and biodegradable synthetic substances are fashioned into the shape of a human organ

6. Begley, "From Human Embryos, Hope for 'Spare Parts'"; Rick Weiss, "Crucial Human Cell Is Isolated."

7. Lessenberry, "Stem Cell Lies."

8. Orr and Hook, "Stem Cell Research: Magical Promise v. Moral Peril," 189–99; Condic, "The Basics of Stem Cells"; Prentice, "Current Science of Regenerative Medicine with Stem Cells," 33–37.

and seeded with stem cells that use this scaffold to form an organ. The stem cells degrade the synthetic scaffold and a newly generated organ is ready for transplantation. Already surgeons have transplanted tissue-engineered bladders and tracheas into human patients.[9] Tissue engineering laboratories have also made engineered livers and kidneys,[10] but they remain unsuitable for human transplantation because of their small size.

My venue is a small Christian university in southern Michigan called Spring Arbor University. There I presented my deliberations on this subject to anyone who might listen, and presentations on embryonic stem cells and regenerative medicine certainly generated a great deal of discussion. This book, therefore, is the result of conversations and emails from my students, mainly, but also my colleagues, staff, and the students' parents. They have asked me many deep and penetrating questions about the personhood of the embryo, stem cells, regenerative medicine, cloning, bioethics, and in vitro fertilization. In this volume I have recapitulated their questions, with some editing and rephrasing for clarity, along with my answers. Many of the questions are reconstructions of conversations, and my answers are also often shortened versions of what usually were much longer conversations.

The first part of the book answers questions about the importance of the stem cell issue and about the process of human development. The next few letters handle basic questions about the nature of stem cells, followed by a series of objections to the personhood of the embryo. In these letters I explain why I am firmly convinced that early human embryos are "one of us" and deserve at least some of the same protections we enjoy.

The middle chapters examine alternative ways to derive embryonic stem cells that might be less objectionable to those of us who take the full human personhood of the embryo seriously. I will try to explain why I find some of them more workable than others. This section also examines cloning and why I think no good can result from human cloning, be it therapeutic or reproductive. All kinds of human cloning should simply be

9. Atala et al., "Tissue-Engineered Autologous Bladders," 1241–46; Atala, "Tissue Engineering of Human Bladder," 81–194; Macchiarini et al., "Clinical Transplantation of a Tissue-Engineered Airway," 2023–30; Chistiakov, "Tissue Engineering and Transplantation," 92.

10. Soto-Gutierrez et al., "Whole-Organ Regenerative Medicine Approach for Liver Replacement," 677–86; Chen et al., "Humanized Mice with Ectopic Artificial Liver Tissues." 11842–47; Joraku et al., "*In vitro* Generation of Three-Dimensional Renal Structures," 129–33; Guimaraes-Souza et al., "*In vitro* Reconstitution of Human Kidney Structures," 3082–90.

banned. This section also examines in vitro fertilization, which is the source of all the excess embryos sitting in cold storage units. While this subject requires a much more in-depth treatment, the lack of proper regulation and oversight of the in vitro fertilization industry has led us to the situation in which we presently find ourselves.

The final chapters survey some of the available adult stem cells and the treatments they presently offer or might be able to offer someday. Stem cells are remarkable entities, and they are changing the face of medicine. Nevertheless, stem cell treatments carry potential risks, as does any medical procedure or treatment, and not every patient will be equally or even successfully treated by them. Furthermore, there are also distinct limits to what stem cells can do or what our present mastery of stem cell biology will allow us to do. Thus while stem cells are extraordinary tools for modern medicine, they are not a panacea for all maladies or a theriac for any poison. They are one of the most incredible agents for healing that modern medicine has yet to see, but they cannot completely remake a human body.

Unfortunately, the incredible potential of stem cells has also led to quack treatments that are untested and that may harm hopeful patients who will spend anything on a potential miracle. This "stem cell tourism" is the result of clinics in India, China, Mexico, Panama, Thailand, and Ukraine that aggressively promote their therapies to foreigners. While many of these treatments are innocuous and unproven at best, some are potentially dangerous. Rather serious complications of stem cell therapies have been documented, including cases of meningitis, a Russian boy who was injected with fetal neural stem cells and subsequently developed brain and spinal tumors, and another case in which a stem cell transplant led to potentially lethal lesions.[11]

On the flip side, legitimate enterprises that offer rigorously tested stem cell–based treatments are being told by the Food and Drug Administration (FDA) that using a patient's own cells is equivalent to administering a drug. The Centeno-Schultz clinic in Broomfield, Colorado, has developed the Regenexx procedures for using a patient's own stem cells to treat orthopedic conditions. Even though Centeno et al. have published multiple studies on their procedures and established the safety and efficacy of them,[12] the FDA has advanced the ludicrous argument that the cells in one's body are drugs,

11. Brown, "Stem Cell Tourism Poses Risks," E121–22.

12. Centeno et al., "Safety and Complications," 81–93; Centeno et al., "Reporting Update," 368–78.

and they get to regulate them.[13] Thus one of the Regenexx procedures, the Regenexx-C procedure that cultures bone marrow–derived stem cells with platelets from circulating blood, has been rendered illegal in the United States by the FDA because they regard the reintroduced cells as drugs. Therefore, this procedure is offered only outside the United States, in the Cayman Islands. While we should eschew untested procedures and the illicit use of stem cell products, we should encourage the careful yet innovative use of stem cells in medicine.

While the stem cell issue is not in the limelight for now, it is not going to go away and will almost certainly come to the fore again in the years and decades ahead. Neither will the strong emotions elicited by this issue go away. Human life and the nature of human life have always been of supreme importance of those who want to protect it and to those who wish to exploit it. When one's definition of a human person threatens someone else's livelihood, it results in enormous battles (recall that the United States fought a civil war over this very question). Thus it is my hope that this book will provide those interested in this topic with something to say, and even more to think about.

13. Regenexx (blog), ""FDA: Your Body is a Drug and We want to Regulate It," February 2, 2012. Online: http://www.regenexx.com/2012/02/fda-your-body-is-a-drug-and-we-want-to-regulate-it.

Acknowledgments

IN ORDER FOR A book to come to fruition, the author must work extremely hard and depend upon the skills of talented people whose contributions play an integral part in the published work. Thus this present work is the fruit of an author who worked his fingers to the bone, but who also relied on some immensely gifted and generous people to produce the finished product you now hold in your hands.

First of all, my wife, Carolyn, who is a gift beyond all measure, kept my home life ordered and sane throughout all the long days and nights of reading, writing, and research. My three daughters, Rachel, Stacy and Emily, likewise have borne their dad's preoccupation with regenerative medicine rather well. To my family I wish to convey my deepest thanks.

Second, I must extend my heartfelt gratitude to Dr. David Hopper of the Spring Arbor University School of Education, Mr. Brian Shaw of the art department, and Mr. Douglass Munn, who worked long hours on the figures to get them into the right format and make them the right size. Gang, I owe you.

Third, no one can do any serious research without a crack interlibrary loan technician, and at Spring Arbor University we have one. Ms. Katherine (Kami) Moyer found all kinds of papers from many various and sundry journals. She has worked tirelessly for me and never complained. Kami, I could not have done this without you. Honorable mention also goes to reference librarians Mr. Robbie Bolton, who was a great help with Chicago Style, and Ms. Karen Parsons, who found several magazine and newspaper articles for me.

Fourth, to my students, fellow Spring Arbor University staff and faculty, and parents, who wrote me some really thoughtful letters, asked great questions, read several drafts of many of these chapters and critiqued them thoroughly (and in some cases mercilessly)—all of you are the reason I teach at Spring Arbor University. You guys are my joy and my crown.

Fifth, to Dr. Charles Morrisey, Emeritus Professor of English at Spring Arbor University, his former colleague Dr. Marsha Daigle-Williamson, and a former student of theirs, Mrs. Jeanette Rick Parker, I express my thanks for proofreading the manuscript. I must also thank Mr. Jacob Martin at Wipf and Stock Publishers for his expert and highly professional copyediting of my manuscript.

Finally, to the triune God who exists as one being in three persons, Father, Son, and Holy Spirit, and is the giver of all life, embryonic and otherwise.

Abbreviations

AFSC	Amniotic fluid stem cell
ANT	Altered nuclear transfer
ART	Artificial reproductive technology
ASC	Adult stem cell
EB	Embryoid bodies
EPC	Endothelial progenitor cell
ESC	Embryonic stem cell
FGF	Fibroblast growth factor
GVHD	Graft-versus-host disease
hAEC	Human amniotic epithelial cell
hCGH	Human chorionic gonadotropin hormone
HSC	Hematopoietic stem cell
ICM	Inner cell mass
iPSC	Induced pluripotent stem cell
IUD	Intra-uterine device
IVF	In vitro fertilization
MSC	Mesenchymal stem cell or mesenchymal stromal cell
NSC	Neural stem cell
NT-ESC	Nuclear transfer embryonic stem cells
OPC	Oligodendrocyte progenitor cell
PGD	Preimplantation genetic diagnosis
SBB	Single blastomere biopsy
SCNT	Somatic cell nuclear transfer (cloning)

Letter #1

Why Should I Care?

Dear Dr. Buratovich,

We have talked about stem cells in class and I must say that the whole thing has me befuddled. What's the big deal? Why should I care? The papers keep running articles about stem cell this and stem cell that, but I have to admit that I just can't get motivated enough to read them and I usually give them a miss. I must admit that I am finding it more than a little difficult to get worked into a lather about the whole thing.

You said that you wanted to hear from us, so here's my earful with a question. My question is this: what is it about embryonic stem cell research that should make me sit up and listen?

Kara B.

Dear Kara,

Why should you care about embryonic stem cell research? If you read the papers, it is clear that embryonic stem cell research discussions make people rather angry. For example, Rick Weiss, the science editor at the *Washington Post*, implied that opponents of embryonic stem cell research are "religious fundamentalists," akin to the Taliban.[1] If that's not strong enough for you, try University of Pennsylvania bioethicist Arthur Caplan, who labeled opponents of human cloning as a "bizarre alliance of antiabortion religious zealots and technophobic neoconservatives along with a smattering of scientifically befuddled antibiotech progressives . . ." Caplan further charged that such people are far more concerned about "cloned embryos in dishes" than "kids who can't walk and grandmothers who can't hold a fork or breathe."[2]

Opponents of embryonic stem cell research can also dish out their share of harsh language. Consider the words of science writer Michael Fumento, who wrote this regarding embryonic stem cell research: "Rightly or wrongly, use of embryonic cells invokes visions of Dr. Josef Mengele and a terrifying slippery slope towards playing around with human life."[3]

Why are these folks so upset at people who disagree with them? I think it comes down to one thing: human life. Human life is something we all care about deeply. This is the one reason why people get worked up about embryonic stem cell research.

Here's the big reason why Christians should care about it. God is the source of life. Only He gives it and only He takes it away. Consider the words of Scripture. In the creation narrative, God breathes life into His creatures and then places a tree of life in the midst of the garden in Eden (Gen 1:30; 2:7, 9). To the Israelites, He said, "I put to death and I bring to life" (Deut 32:39). Nehemiah said that God gives "life to everything, and the multitudes of heaven worship you" (Neh 9:6). Job lamented that "in his hand is the life of every creature and the breath of all mankind" (Job 12:10). The prophet Isaiah wrote, "This is what God the LORD says—he who created the heavens and stretched them out, who spread out the earth and all that comes out of it, who gives breath to its people, and life to those who walk on it" (Isa 42:5). The prophet Ezekiel preached to the dry bones, and

1. Weiss, "Bush Unveils."
2. Caplan, "Attack of the Anti-Cloners."
3. Fumento, "Short on Facts."

the Spirit of the Lord brought them to life (Ezek 37). The prophet Daniel said that God "holds in his hand your life and all your ways" (Dan 5:23).

The New Testament continues this theme. John's Gospel says this about Jesus: "In Him was life, and that life was the light of men" (John 1:4). Believing in Jesus is the difference between having life and not having it (John 3:36). Jesus is called the "Bread of Life" who came down from heaven (John 6:33). Jesus came that "they may have life, and have it to the full" (John 10:10). Knowing the only true God, Jesus, is the source of life that never ends (John 17:3). God is the author of life (Acts 3:15) and gives life to all men (Acts 17:25). Eternal life is the gift of God (Rom 6:23). Those who believe in Jesus have their names written in the "book of life" (Phil 4:3; Rev 20:12). Since the Bible portrays God as the ultimate source of life, if our public policies involve the taking of life or the failure to properly care for life with the resources at our disposal, then our policies are not God-honoring.

Consider how the early Christians put these principles into practice. The Greco-Roman culture into which Christianity was born had little regard for human life and even less for the lives of the weak and vulnerable. Infanticide, infant abandonment, abortion, and suicide were commonplace and even encouraged, as was the barbarism of the gladiatorial games.[4] The response of Christians to this culture was not accommodation, but outright and active opposition. Christian writers wrote against it, but even more telling is that they rescued and raised abandoned babies, some of whom were deformed.[5] The early Christians thought that the more helpless and vulnerable the life, the more deserving it was of compassion and protection.

How does this apply to embryonic stem cell research? Making embryonic stem cells requires the destruction of human embryos. If a human embryo is a human person, then this research requires the deliberate killing of human beings. For the Christian, the destruction of embryos represents the killing of the most vulnerable and helpless in our society. If, however, a human embryo is not a human person, then this research can potentially lead to cures, and stopping this research means that we will slow the development of these cures. If we stop embryonic stem cell research, will people, who might have been cured, die? Maybe, but now you can see why people get angry when it comes to this issue.

4. Schmidt, *Christianity*, 48–74.

5. Ibid., 53.

It gets worse, though. In this country a woman can have an abortion any time during her pregnancy for any or no reason. What's to keep scientists from cloning embryos that are then gestated in volunteer women and later aborted for use in clinical trials? This is called fetal farming; New Jersey has legalized such experiments,[6] and other states have introduced similar legislation. Some scientists want to even use this technology to create designer babies. A group called "transhumanists" wants to remake the human race in their image. The World Transhumanist Association calls this the "post-human species."[7] Should we be concerned? Absolutely. This is nothing short of killing a vulnerable member of the human race, and playing God too. Paying women to have babies just so we can dismember them to use their cells for our own purposes is simple murder. We should be concerned and appalled.

On the subject of cloning, are we comfortable with scientists making embryos in the laboratory just to destroy them? In this case we do not have a woman's choice to consider, we only have embryos that are being made just so they can be killed! Should we be concerned?

On the other hand, can stem cell treatments help sick people? The answer is an unqualified "Yes!" However, we have treatments from stem cell sources other than embryonic stem cells. As it turns out, your body is chock full of stem cells, and scientists have been harvesting them from bone marrow, umbilical cord blood, and other places to treat sick people. Over seventy different conditions have been treated—or at least patients with these conditions have improved, using these "other stem cells."[8]

Where does this leave us? Stem cell research is great. It saves lives and can help lessen human suffering. If you are a Christian, you must be concerned about life. Stem cells can save and extend human lives. That's a huge plus. You should be pro-stem cells and pro-stem cell research.

But what about the embryos that will die to make embryonic stem cells? That's a big minus. Remember, if you are a Christian, you should care about life, even if it is young and immature life. So you should be pro-embryo.

This leaves us in a bit of a conundrum. We want to see stem cell treatments come to the clinic, but we want even the youngest of us to receive the

6. Smith, "Contrary to a Popular Assumption."

7. See http://humanityplus.org.

8. Do No Harm—The Coalition of Americans for Research Ethics, "Fact Sheet." See http://www.stemcellresearch.org/facts/treatments.htm.

legal protection that prevents them from being killed—the same protection that we enjoy. Therefore, we want to see alternatives to embryonic stem cells and we want embryonic stem cell research to move in that direction. However, the papers tend to treat people who oppose embryonic stem cell research to any degree as religious ideologues who want to impose their narrow view of the world on everyone else. Therefore, if you are going to talk about this issue at all, you must do your homework.

This is the scoop on stem cells. Interested in more? Come to class on Wednesday and we'll talk more about it.

Michael Buratovich

Letter #2

Making a Baby

Dear Dr. Buratovich,

Thanks for those stem cell web sites you gave us. They have lots of great information, but reading them is kind of like drinking from a fire hydrant. My big problem is that they use all this jargon like "blastocysts," "blastomeres," "trophectoderm," and "compaction." What on earth does all this mean? I feel like I know more about stem cells but have no clue as to what stage of human development they come from.

I was unable to take your human development class last year. I've heard it's a great class. Is there any way you can give me a primer on human development? It will help me synthesize all this stem cell stuff I've been learning.

No rush on this.

Thanks,

Bobby V.

Dear Bobby,

It is a teacher's dream to see students actually using the suggested resources. Thank you for making my day!

Your request is a very tall order, but I will try my best to summarize what requires a whole semester to actually teach. Here goes nothing.

Since you already know about the "birds and the bees" I will cut to the chase. The preparation for human development starts with an egg that is ovulated by a young woman approximately fourteen days after the end of her last menstrual period. During ovulation, the ovary forcibly expels an egg into the oviduct (or fallopian tube). In vitro fertilization studies have shown that freshly ovulated eggs can only be fertilized up to twenty-four hours after ovulation, after which the egg degenerates. The egg moves down the oviduct where it encounters sperm from the male, and the fusion of the egg and the sperm constitutes the first step of fertilization.

Fertilization is a multistep process that includes several stages. First, the sperm must penetrate the various layers of the egg. Just outside the egg is a thick jelly layer called the zona pellucida, which is a fancy way of saying a clear zone. This jelly layer plays an important role in the early stages of human development. Once the sperm fuses with it, the egg, which was frozen in the middle of cell division, completes cell division and extrudes a tiny clump of unnecessary chromosomes called a polar body. Now the sperm is disassembled within the egg to unveil the sperm pronucleus (another fancy term for a simple thing—the sperm's chromosomes, packaged inside a membrane vesicle). The egg also has its chromosomes packaged in a vesicle called the egg pronucleus. The sperm and egg pronuclei fuse together, and the first cell division then begins.

The fusion of the sperm and egg pronuclei marks the end of fertilization and the existence of the sperm and egg. Originally, the egg produced by the mother's ovaries had half the normal number of chromosomes. By the end of fertilization, the resulting cell has the full complement of twenty-three pairs of chromosomes, half of which came from the father and half from the mother. It is no longer an egg, but a zygote.

Fertilization also kicks the metabolism of the egg into high gear, eventually preparing the zygote for the energy-intensive process of cell division, or cleavage. The first cleavage event occurs about twenty-four to thirty hours after the beginning of fertilization (male embryos actually divide slightly faster than female embryos).[1] Fertilization is also referred to as conception. *Conception*, however, is an inexact term, since some use it

1. Pergament et al., "Sexual Differentiation," 1730–32.

to refer to implantation of the embryo into the uterus and others to refer to fertilization. For that reason, conception is not a term that is used in developmental biology, except in the term *conceptional age*, which refers to the true age of the unborn baby. Conception, by which I mean the process of fertilization, is a multistep process, and its completion produces an entity that now begins the seamless and continuous process of human life, which includes embryonic, fetal, and postnatal development. Because this continuous process defies demarcation, it is most accurate to define conception as that event which brings a human being into existence.

Fertilization occurs within the oviduct, as do the first cell divisions or cleavages. Cleavages, or repeated cell divisions of the embryo, divide the zygote into two cells, then twelve to eighteen hours later into four cells, and within eighteen to twenty-four hours into eight cells (fig. 2.1). The embryo, at this time, does not increase in size, but is only divided into smaller and smaller cells, enclosed within the zona pellucida. By three days after fertilization, the embryo consists of six to twelve cells, and by four days, sixteen to thirty-two cells. On the third day of development, the embryo also begins to wean itself from its dependence on the materials initially stock piled into the egg and establishes its own gene expression program.[2] An embryo with twelve to thirty-two cells is called a morula.

After the eight-cell stage, somewhere around the twelve to sixteen-cell stage, human embryos undergo a process called compaction. Before compaction, the cells of the embryo, which developmental biologists call blastomeres, do not possess tight connections with each other. At compaction, the outer blastomeres bind tightly to each other and force particular blastomeres inside the ball of cells, while the others remain on the outside. Compaction generates two populations of cells in the embryo: cells on the outside and cells on the inside. The outer cells eventually flatten and become the cells of the trophoblast. The trophoblast eventually develops into the placenta. Inside the embryo, the inner cells round up and form the embryoblast, or inner cell mass (ICM). The ICM cells will make the embryo proper and add a few elements to the placenta.

By the fourth day of development the trophoblast cells begin to pump salts into the interior of the embryo, and this causes water to follow (fig. 2.1). This results in the swelling of the embryo into a sphere with a cavity inside it. Now the embryo is called a blastocyst, and its ICM cells are

2. This phenomenon is known as embryonic genome activation or EGA. See Dobson et al., "Transcriptome Through Day 3," 1461–70; Wong et al., "Human Embryos Before Embryonic Genome Activation," 1115–21.

clumped together at one end of the embryo (the embryonic pole). At this time, the embryo usually completes its journey from the oviduct to the uterus. The trophoblast cells also make a protein called early pregnancy factor that finds its way into the mother's bloodstream. This protein is found quite early after fertilization and is the basis for pregnancy tests applied during the first ten days of development.

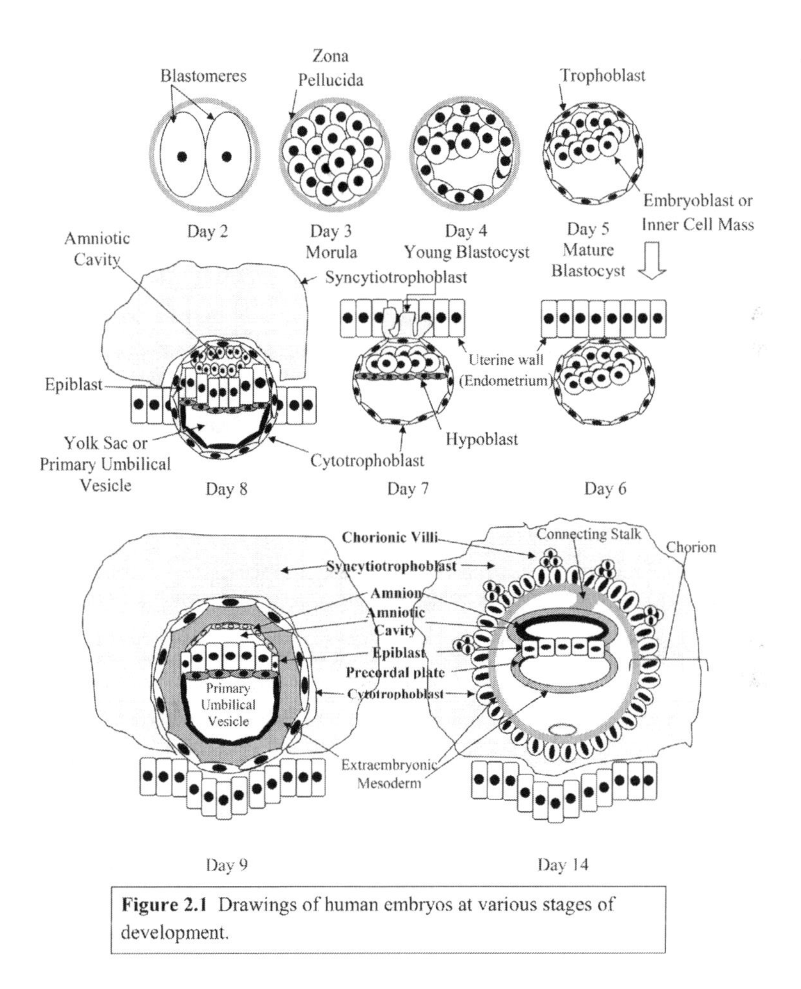

Figure 2.1 Drawings of human embryos at various stages of development.

Five days after fertilization, the embryo bores a hole in the zona pellucida and escapes from its early protective membrane. This is called hatching of the blastocyst. The embryo spends approximately two days in the uterus before it implants into the uterine wall.

The five-day embryo (containing approximately 150 cells) is the entity sought after by embryonic stem cell researchers. The ICM cells are those that will form embryonic stem cell cultures. To make embryonic stem cell cultures, researchers use either mechanical or chemical means to disassemble the trophoblast, and then culture the isolated ICM cells under specific conditions. If the ICM cells grow, they can become an embryonic stem cell line.

Implantation begins on the sixth day, and the embryo is only competent to implant during a short window of time, after which it loses its ability to implant and dies. The trophoblast adheres to the surface of the uterus, and this contact induces the trophoblast cells to divide. Some of these dividing trophoblast cells fuse together and become a kind of amoeba that rapidly moves into the uterine wall, digesting as it goes (syncytiotrophoblast). The trophoblast cells that do not fuse (cytotrophoblast) remain in place and divide there. Between days six and nine, the embryo sinks below the surface of the uterine wall (fig. 2.1). During this time, the ICM cells organize into a sheet of tall cells known as the epiblast. The cells of the epiblast divide and form two structures above and below it. Below the epiblast, a layer of cube-shaped cells (hypoblast) forms and spreads downward toward the other end of the embryo. Above the epiblast, a small layer of cells peels off and grows around the embryonic pole, forming a small cavity called the amniotic cavity. The end result is a vesicle above (amniotic cavity) the single-cell-thick epiblast, a single-cell-thick hypoblast beneath the epiblast, and a vesicle beneath the epiblast called the primary yolk sac or primary umbilical vesicle that is continuous with the hypoblast. The cells of the outer layer of the primary umbilical vesicle make an extensive gel-like material between the vesicle and the cytotrophoblast known as the extra-embryonic mesoderm. This layer fills all the space between the amnion, primary umbilical vesicle, and cytotrophoblast.

By day fourteen, the syncytiotrophoblast has digested blood vessels in the uterine wall and is filled with small pools of blood that provide oxygen for the growing embryo. Also the primary umbilical vesicle has shrunk and pulled away from the cytotrophoblast. The amnion and its cavity, the epiblast, and the primary umbilical vesicle are suspended within the cytotrophoblast by a connecting stalk. The cytotrophoblast has begun to form small bumps called primary chorionic villi, which will eventually form the blood vessels of the placenta and anchor the placenta to the uterine wall (fig. 2.1). The hypoblast has a slight thickening at one end of the embryo, and this structure, the prechordal plate, marks the site of future head formation.

After fourteen days of life, the embryo begins a remarkable series of rearrangements known as gastrulation. Gastrulation transforms the single-cell-thick epiblast into a three-layered structure. In the middle of the epiblast, right at the back of it, a thickened ridge forms called the primitive streak. The formation of the primitive streak marks the future dorsal side and backside of the baby (your spine is on the dorsal side of your body and your stomach on the ventral side; your head is the cranial end and your rump is your caudal end). The front of the primitive streak enlarges into a structure called the primitive node, and the primitive node points directly toward the future head (fig. 2.2). Signals within the embryo also tell it which is the left side and which is the right side. Because particular events during development, like the rotation of the heart and the digestive system, occur in particular directions, it is crucial that the embryo "knows" which side is which. This is also about the time when the mother misses her first menstrual period.

Cells begin to pour through the primitive streak, and take up specific positions underneath the epiblast. The primitive streak elongates to become half as long as the epiblast. The first cells that move into the embryo insert themselves into the hypoblast and become known as endoderm. Endoderm will form the respiratory and digestive systems that help us breathe and digest food. The next group of cells that moves into the embryo become mesoderm. Smooth muscle, connective tissues and blood vessels, the heart, muscles, and the reproductive and excretory systems are all formed by mesoderm. Those cells that remain on the dorsal side are known as ectoderm, and these cells make the skin, nervous system, and sensory organs like the eyes. Together, the ectoderm, mesoderm, and endoderm constitute the embryonic germ layers, and these three layers will form all the tissues in the baby.

By the seventeenth day, the primitive streak begins to retract and in its wake induces the formation of a flat, thickened plate called the neural plate. This is the beginning of the nervous system, and the edges of the neural plate fold and rise as a crease forms along the center of the plate. The neural plate folds in half along this lengthwise crease, which causes the two folds to meet above the middle of the plate, where they fuse to form a tube (fig. 2.2). This tube, the neural tube, is the beginning of the spinal cord, and the front portion of it inflates to form the brain. Beneath the neural plate, cells condense to form a stiff rod called the notochord, and this structure is the beginning of the future vertebral column that houses the spinal cord.

The notochord then induces nearby mesoderm to clump and form somites. Somites will form the muscles of the back and the bones of the vertebral column.

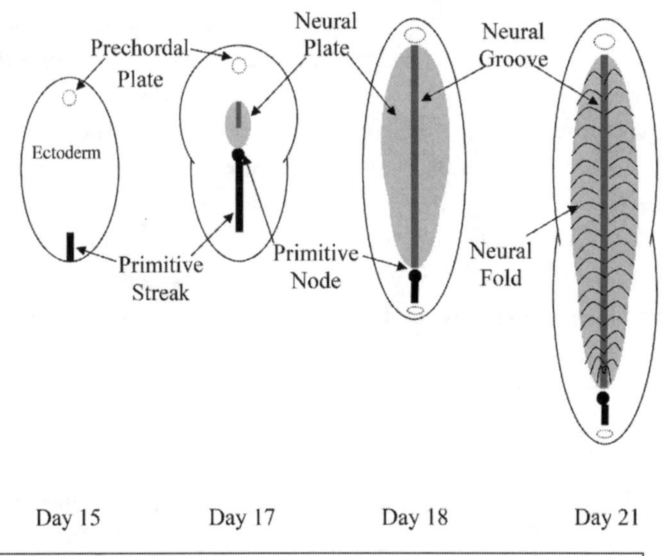

Figure 2.2 These drawings illustrate "neurulation" or the making of the nervous system. These figures depict the epiblast if someone was looking down on it with the amnion and other covering structures removed. The neural plate rises and becomes neural folds. These neural folds on either side of the neural groove move toward each other and fuse to form a hollow tube called the neural tube. The neural tube is the embryonic spinal cord. Neurulation is complete during the fourth week of life.

From these tissues, various embryonic organs form. On day twenty, the thyroid gland begins to form, and by day twenty-two, the heart begins to beat. By the next day the beginnings of the eye and ear are present. On day twenty-four, the pharyngeal arches appear; these structures will form a good portion of the facial structures, middle ear, tonsils, parathyroid glands, and thymus. On day twenty-six, the beginnings of the arms form as a bump called the limb bud. At twenty-seven days, the forebrain forms, and one day later, the embryo is about four millimeters long. On day thirty-three, the hand starts to form (hand plate), and one day later the foot begins its formation (foot plate). On this same day, the distinct subdivisions of the brain are apparent. By day thirty-five the embryo is eight millimeters long.

By thirty-eight days, the upper lip and nose are formed. By forty days, the external ear and the fingers begin to form. By the forty-third day of life the embryo is sixteen millimeters long, and by the forty-fourth day the eyelids are forming. By fifty days of life, the arms are bent at the elbows and the fingers are webbed. By the fifty-third day the external genitals begin to form. At fifty-six days, the embryo is thirty millimeters long, and this marks the end of the embryonic period.

At the end of the embryonic period of human development, all the organs are in place, but they are not yet mature. The fetal period is characterized by growth and detailed elaboration of these structures. At nine weeks of life the eyelids are fused. The intestines grow so fast that they must move into the umbilical cord by the sixth week, where they remain until the eleventh week, at which point they return to the abdomen. Between nine and twelve weeks urine formation begins. The fetus recycles much of the amniotic fluid in which it is suspended by swallowing it and re-excreting it in urine, after which the placenta filters the metabolic wastes. The sex of the fetus is clearly distinguishable by twelve weeks of life. By seventeen weeks, the mother can typically feel fetal movements, and by week twenty the eyebrows and hair are visible. Rapid eye movements begin by week twenty-one, and the "blink-startle response" begins by week twenty-two. Even though a fetus is viable at twenty-two weeks of life, her chances for survival are poor. Fingernails are in place by the twenty-fourth week. By twenty-six weeks, the baby's eyes open, and by thirty weeks the pupils of the eye respond to light. During the last weeks in the womb the baby gains fat.

The expected date of delivery is 266 days after fertilization, or about thirty-eight weeks. Some 12 percent of babies are born one to two weeks after the expected delivery date.

This is a highly abbreviated description of human development, but it explains those events that are important to understand when it comes to embryonic stem cell research. Remember that embryonic stem cells are thought to be the most useful cell for therapeutic purposes because they can form cells from any of the three embryonic germ layers; however, adult stem cells are much more versatile than was previously thought.

I hope this helps. If you have any questions, please feel free to e-mail me again or ask me in class.

Cheers,
Michael Buratovich

Letter #3

Stem Cells 101

Dear Dr. Buratovich,

I am a student in the general science class for elementary school teachers. Our instructor assigned a writing project that requires us to research a contemporary issue in science and write a research paper on it. Since I suffer from a chronic disease (lupus), I have decided to do stem cells for my paper. Several biology majors on my dorm floor told me to go to you, since you know a great deal about this issue.

I came by your office, but you were in lab at the time. Therefore, I thought I would do this by e-mail. Can you please tell me what stem cells are? What are the issues surrounding them and how might they help someone like me who has a chronic disease?

Thank you for your time.

Molly S.

Dear Molly,

Thank you for your note. I feel kind of honored that my students think I have something to offer on this subject. Yes, I have done a fair amount of reading on this subject, and I would love to help you.

So where should we begin? Why don't we start with what stem cells are? Stem cells are special cells that are found in the bodies of multicellular organisms and that have specific characteristics. Here are the basic attributes of stem cells:

1. Stem cells are primal cells, which is to say that they are immature. They are undifferentiated, or have yet to become a specific type of cell. Stem cells also have the capacity to maintain this primal developmental status throughout their lifetime.
2. Stem cells have the ability to self-renew. They are capable of continuous cell division and self-maintenance, which is to say that stem cells can divide and make a carbon copy of themselves, and that carbon copy commits to becoming some other type of cell while the original cell remains a stem cell.
3. Stem cells have an unlimited ability to divide and make more cells. Most cells can only divide a limited number of times before they lose their ability to divide. Stem cells, however, possess an unlimited proliferative capacity.[1]

Stem cells can replace dead and dying cells, or repair an organ if it is injured. Tissue-specific stem cells can help maintain the health of the organ, or help organs grow if they need to.[2]

A key feature of stem cells is this: they are not all equal. Some stem cells can make every type of cell in the embryo or the adult. The cells of the very early embryo are good examples of such cells, and we call such cells *totipotent*. Other stem cells can make any cell in the adult body, but not all types of cells in the embryo. For example, embryonic stem cells from mouse embryos can form all the cells in the adult mouse body, but not all the cells in the embryo. These cells are termed *pluripotent*. Some stem cells can only form a subset of the cells found in the adult body. Stem cells like

1. Bryant and Schwartz, *Fundamentals*, 10. Note that this stem cell–specific ability—the ability to divide indefinitely—is very heavily regulated in order to prevent overproliferation, which is characteristic of cancer cells. See Falzacappa et al., "Regulation of Self-Renewal in Normal and Cancer Stem Cells," 3559–72.

2. Kiessling and Anderson, *Human Embryonic Stem Cells*, 2–3.

these are found in our tissues and help those tissues regenerate and maintain their health and size. Such stem cells are called *multipotent*. Finally, certain stem cells only form one type of cell and these are called *unipotent* stem cells. Satellite cells in our muscles are stem cells that only form muscle, and represent one example of a unipotent stem cell.

There are essentially five different types of stem cells. Each has particular characteristics and is found in particular places.

A. Adult or Somatic Stem Cells

Adult stem cells (ASCs) are formed during the fetal stage of development. As a rule, ASCs are multipotent, but their *plasticity*, or ability to differentiate into different types of cells, is much greater than previously thought. Because they are found in children and adults, the term *adult stem cells* is a misnomer, and *somatic stem cells* would certainly be a more accurate term for these cells. However, since adult stem cells are so well known, introducing a new phrase would merely confuse people.

Bone marrow probably represents one of the most important examples of the therapeutic uses of ASCs. Bone marrow contains a veritable cornucopia of stem cells, but perhaps the most important of these is the blood-making or *hematopoietic* stem cell (HSC). HSCs divide to form all of the different cells found in blood, including the red blood cells that carry oxygen to our tissues, organs and cells, white blood cells that generally protect us from invaders, platelets that help clot our blood when we get cut, and lymphocytes, which are responsible for *adaptive immunity*, whereby our immune system actually remembers infections that happened in the past (fig. 3.1).

The stem cells in bone marrow can also be used to cure other people who suffer from bone marrow-based diseases. From 1955 to 1962, E. Donnell Thomas transferred bone marrow from healthy dogs to sick ones, and showed that if the tissue types of the donor and recipient dogs were properly matched, the sick dogs recovered and got better. In 1963, Thomas began to apply the principles he had worked out in dogs to human patients, and by the early 1970s, patients with leukemias and other bone marrow diseases were being cured with bone marrow transplantations. In 1990, Thomas won the Nobel Prize in Medicine for his pioneering work in bone marrow transplantation. Bone marrow transplantation is a stem cell treatment that has saved thousands of lives to date.

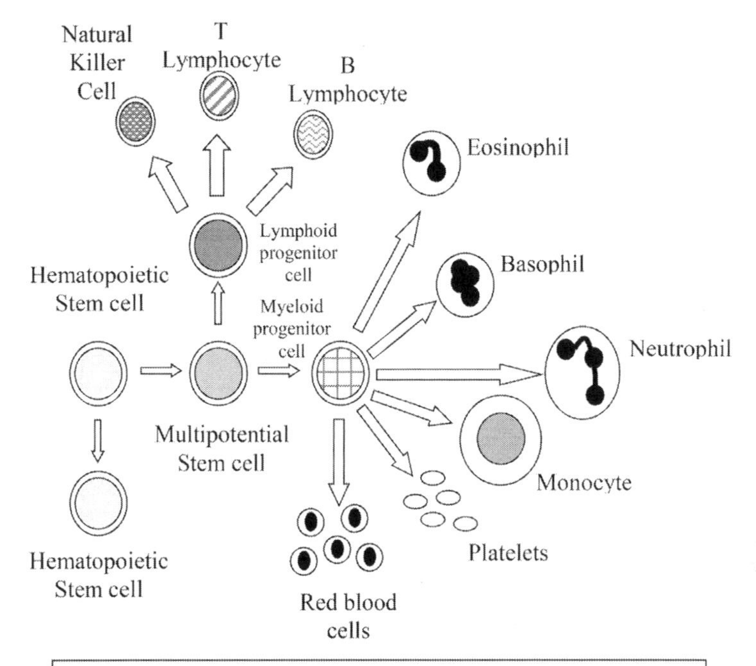

Figure 3.1 Hematopoietic stem cells are found in bone marrow. This very heavily-studied stem cell gives rise to red blood cells and all the other cells regularly found in our bloodstreams, including those white blood cells that play such a key role in innate and adaptive immunity.

Can HSCs from bone marrow differentiate into something other than blood cells? The answer is definitely yes.[3] Laboratory studies have shown that cultured bone marrow–based stem cells can be coaxed into forming more than just blood cells.[4] However, the real clincher came with studies of women who received bone marrow transplants from male donors; following the transplants, it was found that these women had cells with Y chromosomes in places other than their bone marrow. Remember that female humans have two X chromosomes and male humans have an X and a Y chromosome. The fact that these women had cells with Y chromosomes

3. Andreas Hüttmann et al., "Bone Marrow-Derived Stem Cells," 599–604.
4. Stewart Sell, "Stem Cells," 1–18.

in their muscles, livers,[5] pancreases, blood vessels, hearts,[6] brains,[7] lungs, kidneys, and retinas strongly suggests that bone marrow-based stem cells can do more than just make blood cells.

ASCs like those in bone marrow and many others provide the resources for a field of medicine called regenerative medicine. Physicians can potentially remove ACSs from one tissue that is not sick and transplant it to a tissue that is sick so that the newly-introduced stem cells can heal the ailing tissue. Such a protocol provides the means for sick people to be treated with their own stem cells. This type of treatment is called an *autologous transplantation*, and this strategy has been utilized to treat several diseases.[8] In fact, scientists have used autologous HSC transplantations to treat patients who, like you, struggle with lupus.[9] A different treatment strategy is to use stem cells from a tissue-matched donor who kindly provides their stem cells to treat you. This type of treatment is called an *allogeneic transplantation*, and it has also been used to treat several different diseases.[10]

B. Fetal Tissue Stem Cells

Fetal organs are all in place by three months of development, and at this time the fetus weighs about one ounce and is only a couple of inches long. During the remaining six months of in utero development the organs will enlarge and eventually become functionally independent of the placenta. The rapid growth of the fetal organs is due to a robust population of fetal stem cells that reside in these organs. Fetal stem cells also tend to be multipotent, however they typically can form more types of cells than their ASC counterparts.[11]

5. Neil D. Theise et al., "Liver from Bone Marrow in Humans," 11–16.

6. J. Thiele et al., "Regeneration of Heart Muscle Tissue," 201–9.

7. Mezey et al., "Transplanted Bone Marrow," 1364–69.

8. For examples, see Burt et al., "Clinical Applications," 925–35; Rabusin et al., "Immunoablation," 81–85.

9. Burt et al., "Nonmyeloablative Hematopoietic Stem Cell Transplantation," 527–35; Tyndall and Uccelli, "Multipotent Mesenchymal Stromal Cells," 821–28; Zeher et al., "Autologous Haemopoietic Stem Cell Transplantation," 1193–201.

10. Amrolia et al., "Nonmyeloablative Stem Cell Transplantation," 1239–46; Khouri, "Allogeneic Stem Cell Transplantation," 271–77; Uccelli et al., "Mesenchymal Stem Cells," 649–56.

11. For examples, see Srikanth et al., "Fetal Mesenchymal Stem Cells Differentiate into All Three Germ Layers," 26–33; Jones et al., "Differences in First Compared to Third Trimester Human Fetal Stem Cells," e43395.

Ethical problems arise with the use of fetal stem cells, because harvesting them requires the dismembering of the fetus. Having said that, some scientists have done research with fetal stem cells isolated from stillborn infants,[12] which indicates that not all fetal stem cell research is done on the back of abortions. Still, the bulk of it is done with tissue harvested from the bodies of aborted fetuses.

In the late 1980s and early 90s, patients with Parkinson's disease, a condition that reduces people's ability to control their movements, were treated with transplantations of human fetal brain cells. While these patients seemed to initially improve, further examination showed that these treatments conferred little to no benefit.[13] Post-mortem examination of the brains of patients who had received the fetal brain cell transplants also revealed something quite interesting. Although implanted fetal brain cells survived and integrated into the patient's brain, in several cases, the transplanted cells showed the same pathology as the surrounding tissue. In other words, the brain tissue of the Parkinson's patient transferred the disease to the implanted fetal tissue.[14] Fetal tissue transplants did not provide satisfactory treatment in this case.

C. Embryonic Stem Cells

Embryonic stem cells (ESCs) are made from human embryos. In most laboratories, embryos for research are made by means of in vitro fertilization at fertility clinics. Some of the embryos that are made in excess of the reproductive needs of their clients are donated by fertility clinics, with the consent of the couples whose eggs and sperm were used to make the embryos.

After fertilization, human embryos undergo cell divisions without any growth in total volume, which means that the cells of the embryo get smaller with each cell division. By four to five days after fertilization, the embryo consists of 50–150 cells and has rearranged itself into a hollow sphere called

12. Klassen et al., "Isolation of Retinal Progenitor Cells," 334–43.

13. Freed et al., "Transplantation of Dopamine Neurons," 710–19; Freed et al., "Patients with Parkinson's Disease," III44–46; Robinson, "Fetal Transplants for Parkinson's Disease," 69.

14. Kordower et al., "Lewy Body–Like Pathology," 504–6; Li et al., "Lewy Bodies in Grafted Neurons," 501–3; Mendez et al., "Dopamine Neurons Survive without Pathology for 14 Years," 507–9; Ahn et al., "Neighboring Tissue and Pathology in a Fetal Transplant for Parkinson's Disease," 49–59.

a blastocyst. The blastocyst contains two types of cells. On the outside are trophoblast cells that will make the placenta and on the inside are cells of the inner cell mass (ICM), which will make the embryo itself.

Human ESCs are harvested from blastocyst-stage embryos. The trophoblast cells are removed by a variety of means and the ICM cells are cultured on a plastic dish coated with a layer of feeder cells. These feeder cells are particular mouse or human skin cells called *fibroblasts* that make chemicals that help the ICM cells survive and prevent them from differentiating.[15] The ICM cells grow and flatten into a compact colony of ESCs. These colonies are mechanically removed and replated multiple times. If they grow and survive, which is by no means guaranteed, they will form a stable ESC *line* (fig. 3.2).

In culture, human ESCs, like those from other mammalian species, must be grown on layers of feeder cells to survive. While mouse fibroblasts have been used for feeder cells, human fibroblasts have been used and do work.[16] If ESCs are going to be used for human regenerative medicine, they must be produced without animal products, since potentially dangerous animal viruses can pass from animal cells to human cells.[17] If grown under the right conditions, ESCs can divide indefinitely without differentiating. However, under other conditions, ESCs can differentiate and form essentially any cell in our bodies.[18] If undifferentiated ESCs are transplanted into special laboratory animals, they form an unusual type of tumor called a teratoma.[19] Teratomas contain a mixture of a wide variety of cells. The ability of ESCs to form teratomas illustrates their pluripotency. However, this ability is also one of the drawbacks of ESCs when it comes to using them for therapies, since transplanting ESCs into a patient might cause cancer.[20]

15. Thomson et al., "Embryonic Stem Cells Lines," 1145–47; Cowan et al., "Derivation of Embryonic Stem-Cell Lines," 1353–56.

16. Amit et al., "Clonally Derived Human Embryonic Stem Cell Lines," 271–78.

17. Human embryonic stem cell lines have been made in animal product-free environments. Such ESC lines are known as "Xeno-free" lines. See Stephenson et al., "Human Embryonic Stem Cell Lines in an Animal Product–Free Environment," 1366–81; Tannenbaum et al., "Derivation of Xeno-Free and GMP-Grade Human Embryonic Stem Cells," e35325.

18. Bodnar et al., "Undifferentiated Human Embryonic Stem Cells," 243–53.

19. Trounson, "Stem Cells, Plasticity and Cancer," 2763–68.

20. Földes et al., "Cardiomyocytes from Embryonic Stem Cells," 1473–83.

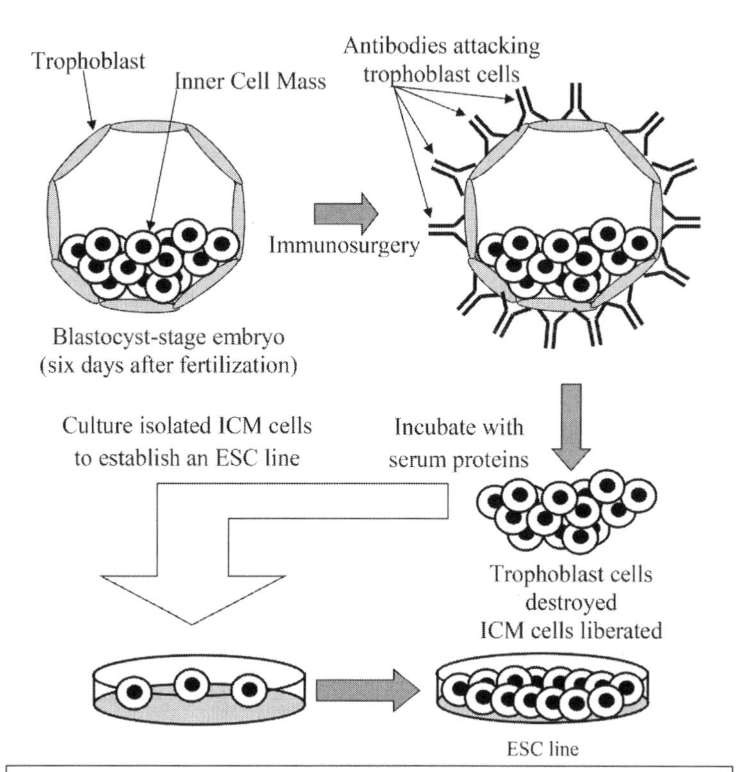

Figure 3.2 ESC derivation from late-blastocyst-stage embryos requires the incubation of the embryos with specific blood-based glycoproteins called antibodies. Antibodies are proteins that specifically bind particular proteins, and these particular antibodies bind cell surface proteins that are found on trophoblast cells. Bound antibodies, in combination with blood-based proteins called complement proteins, team up to bore holes into the surfaces of these cells and kill them. This frees the ICM cells so the researcher can culture them in an artificial culture system to make an ESC line.

If undifferentiated ESCs are removed from feeder layers and transferred into a culture that contains serum (the liquid part of blood minus the clotting factors), they clump together to form embryoid bodies (EBs). Within the EBs, all types of cells form in a random manner. However, scientists have shown that particular conditions can direct the differentiation of the ESCs into specific types of cells. These procedures, however, usually only enrich for particular types of cells, and further types of isolation are usually necessary. If particular types of cells can be generated from ESC cultures, then perhaps they can be used to regenerate damaged or dying organs.

The controversy surrounding ESCs centers on one particular fact: making ESC lines requires the death of human embryos. When people have trouble with ESC research, it usually comes down to that one issue.[21]

D. Parthenote Stem Cells

Isolated eggs can be artificially activated with chemicals or electrical shocks. This activation without the benefit of fertilization will cause the egg to divide and even develop to some extent. This is called *parthenogenesis*, and such artificially activated embryos can form blastocysts from which stem cells can be made,[22] but die during the embryonic stage of development.[23] Such stem cells are called *parthenote stem cells* (pES), and they have been isolated from mice, nonhuman primates, and humans.

Scientists have used artificially activated eggs to make tissue-matched pESs from mice[24] and humans.[25] However, pES cells lack the ability to form particular cell types,[26] which limits their clinical usefulness.[27] Secondly, there are indications that pESs and any cells made from them routinely possess abnormal numbers of chromosomes, which makes them prone to die or grow uncontrollably and cause tumors. Such characteristics could call into question the safety and potential clinical usefulness of pESs.[28]

21. Genetic engineering techniques can be used to convert adult cells into stem cells that greatly resemble embryonic stem cells known as induced pluripotent stem cells. Because of their similarity to embryonic stem cells, they are not discussed further here.

22. Vrana et al., "Nonhuman Primate Parthenogenetic Stem Cells," 11911–16.

23. Hauzmann and Papp, "Conception without the Development of a Human Being," 175–77.

24. Kim et al, "Histocompatible Embryonic Stem Cells by Parthenogenesis," 482–86.

25. Revazova et al., "Stem Cell Lines from Human Parthenogenetic Blastocysts," 432–49; Revazova et al., "HLA Homologous Stem Cell Lines from Human Parthenogenetic Blastocysts," 1–14.

26. Nagy et al., "Parthenogenetic Cells in Chimeras," 67–71; Paldi et al., "Parthenogenetic Mouse Aggregation Chimeras," 115–18; Fundele et al., "Elimination of Parthenogenetic Cells," 29–35; Fundele et al., "Selection Against Parthenogenetic Cells," 203–11; Allen et al., "Imprinting in Parthenogenetic Embryonic Stem Cells," 1473–82.

27. Lengerke et al., "Histocompatible Parthenogenetic Embryonic Stem Cells," 209–18.

28. Brevini et al., "Unstable Equilibrium Between Pluripotency and Malignant Transformation," 206–12.

These concerns might be overblown, since some stem cell scientists have successfully produced functional, stable nerve cells from human pESs, which suggests that these cells could someday play a central role in certain regenerative therapies.[29] Also, pESs can serve as model systems for laboratory research.[30]

E. Nuclear Transplant Stem Cells

Isolated eggs can have their chromosomes removed by a process called enucleation, and replaced with chromosomes from another cell. This procedure is called *somatic cell nuclear transfer* (SCNT). SCNT forms an embryo that is genetically identical to the cell that donated the chromosomes. In other words, the cell that donated the chromosomes for SCNT has been cloned. The word clone comes from the Greek word *klon*, which means twig, which refers to the fact that a twig can, if planted, give rise to a genetically identical tree. Cloned embryos can grow to the blastocyst stage and ESCs can be made from them. Such stem cells are called nuclear transplant stem cells (NT-ESC). NT-ESCs seem to have all the capabilities of ESCs made from non-cloned embryos.[31]

Scientists hope that NT-ESCs might be used to prevent the patient's immune system from rejecting implanted cells. If differentiated cells are implanted into patients to heal them, the patient's immune system will reject any cells that are not sufficiently similar to the patient's cells. By making cloned tissues to replace dead tissues, scientists hope to circumvent this problem.

The main issue with cloned embryos is that once again, a human embryo is being brought into existence only to be destroyed. This is the main problem people have with the production of NT-ESCs.

The ability to replace damaged cells with new ones is an extremely exciting possibility. For this reason stem cells and regenerative medicine are some of the most exciting fields in science today. I hope this short

29. Isaev et al., "Differentiation of Human Parthenogenetic Stem Cells into Neural Lineages," 37–45.

30. Ahmad et al., "Functional Neuronal Cells Generated by Human Parthenogenetic Stem Cells," e4125.

31. Munsie et al., "Embryonic Stem Cells from Adult Mouse Somatic Cell Nuclei," 989–92; Wakayama et al., "Embryonic Stem Cell Lines from Adult Somatic Cells," 740–43; Brambrink et al., "ES Cells from Cloned and Fertilized Blastocysts are Indistinguishable," 933–38.

introduction to the field gives you some appreciation of it, and the concerns that it creates as well.

Sincerely yours,
Michael Buratovich

LETTER #4

Embryos and the Early Church

Dr. Buratovich,

My mom read this book that said, for most of its history, the Catholic Church considered the unborn nonhuman until the mother felt it move. Only then did it receive a human status. Before that time, it was unformed or something like that. I told her I didn't believe it, so she loaned me the book and there it is in black and white. What's up with this? Am I being sold a bill of goods?

Rachel K.

Letter #4

Dear Rachel,

By the way, great job at the tennis match last week. You were on fire! I did not know you could serve like that. Well done.

With respect to your question, it is certainly true that some scholars have argued that the Catholic Church did not regard the embryo as a human person for most of its recorded history. They depend on statements by church intellectuals and leaders like Augustine, Aquinas, and Pope Gregory XIV, who, these advocates say, endorsed Aristotle's concept that a new person does not exist until *quickening*, which is the moment the mother first notices movements in her womb. These assertions come from the historical analysis of an Anglican scholar named Gordon R. Dunstan, who claimed that absolute protection for the embryo in Christian thinking is a creation of the nineteenth century.[1] This conclusion does not wash and here's why.

Scripture mentions the unborn several times. God denounces the killing of unborn children in Amos 1:13, where the prophet Amos condemns the Ammonites "because they ripped open expectant mothers in Gilead." Old Testament prophets state specifically that God forms people in the womb (Jer 1:4–5; Isa 44:1–2), which is reinforced by other verses elsewhere (Job 10:8–10; Pss 22:9–10; 51:7; 119:73; 139:13–16). The New Testament affirms this (Gal 1:15), and the Gospel of Luke also records the unborn John the Baptist recognizing the voice of Mary as the mother of His Lord (Luke 1:39–44). In fact, Luke uses the same word (*bréphos*) to describe an unborn baby and a baby who has been born (see Luke 1:41 and 18:15). There might be a possible condemnation of abortion in Gal 5:20, and Rev 9:21, 18:23, 21:8, and 22:15. The Greek word translated as "sorcery" (*pharmakeía*) in these verses was used by other ancient authors to designate the drugs given to produce an abortion.[2] Thus, Scripture teaches that God is intimately interested in the unborn, and may even contain a direct condemnation of abortion.

However, Scripture uses nonscientific language and, unfortunately, does not directly address the point during development when the unborn becomes a human person. None of the above passages tell us when during development personhood arises.

However, because abortion and infanticide were very common in the Greco-Roman world,[3] which served as the setting for the birth of Christi-

1. Dunstan, "Human Embryo in the Western Moral Tradition," 40.

2. Gorman, *Abortion and the Early Church*, 48.

3. Jones, *Soul of the Embryo*, 31–42.

anity, the early church must have had an opinion on the subject. How did they understand the words of Jesus with regard to the unborn?

On this question the historical data are pretty clear: the early church condemned abortion and infanticide with a unified and unwavering voice. Consider the following facts:

- The earliest specific witness to the Christian ethical tradition regarding the unborn comes from the *Didache*, otherwise known as the *Teaching of the Twelve Apostles*. This work dates to the late first century or early second century AD. In its explanation of the second great commandment to "love your neighbor as yourself," it launches into several "thou shalt not" statements that are modeled on the Ten Commandments. One of these statements is "thou shalt not murder a child by abortion nor kill them when born" (2.2).[4]
- The early second-century *Letter of Barnabas*, which is more of a theological tract on Christian living and thinking, makes the same statement (19.10).[5] The unborn is considered a neighbor, and abortion is a failure to show love to a neighbor. Both documents make no distinction between a formed or unformed fetus, and instead forbid abortion at any stage of development.
- The early second-century work *The Apocalypse of Peter* (v. 25), and the late fourth-century work *The Vision of Paul* (v. 40), describe the threat of divine judgment for having an abortion, and portray men and women having to face the children they had aborted.[6] These documents paint abortion as the deliberate murder of a human being, regardless of the stage of the pregnancy.
- When a vicious rumor about Christians engaging in child sacrifices began to circulate in the second century, Christian leaders used their stand against abortion to show that this accusation was ludicrous. As an example, in AD 177 the Christian leader Athenagoras wrote a letter to the Roman Emperor Marcus Aurelius and noted that since Christians viewed abortion as murder, there was no way they would

4. Lightfoot, *Apostolic Fathers*, 123–24.

5. *Epistle of Barnabas*.

6. *The Apocalypse of Peter*, in *The Ante-Nicene Fathers* (hereafter cited as *ANF*), 9:141–47; *The Vision of Paul* (*ANF* 9:149–51).

condone or practice child sacrifices (35.6).[7] Twenty years later, the Christian apologist Tertullian made a similar argument (9.8).[8]

- In the early third century, Christian apologist Minucius Felix pointed out that it was pagans and not Christians who treated the unborn with indifference: "And I see that you at one time expose your begotten children to wild beasts and to birds; at another, that you crush them when strangled with a miserable kind of death. There are some women who, by drinking medical preparations, extinguish the source of the future man in their very bowels" (30.2).[9]
- At the turn of the fourth century, Lactantius (260–330) wrote that Christians were forbidden to kill in not only ways that were illegal and socially unacceptable, but also in ways that were tolerated and esteemed in the pagan world. This was the case with the strangulation of infants and the killing of unborn children who had never seen the light of day (6.20).[10]
- Various sermons and letters from Christian leaders show that when particular Christians failed to live up to their high moral principles and participated in abortions, they were met with condemnation. For example, when Cyprian of Carthage (ca. 200–258) accused the schismatic priest Novatus of inducing a miscarriage when he kicked his wife with his heel, he called the dead offspring "the fruit of a father's murder" (48.2).[11] Similarly, Hippolytus (170–236) accused Callistus (then bishop of Rome) of recognizing mismatched marriages between high-born women and men of low social status that resulted in abortions (9.7): "Whence women, reputed believers, began to resort to drugs for producing sterility, and to gird themselves round,[12] so to expel what was being conceived on account of their not wishing to

7. Athenagoras, *A Plea Regarding Christians*, in Richardson, *Early Christian Fathers*, 1:290–340.

8. *Apology* (*ANF* 3:25–26).

9. Municicius Felix, *Octavius* (*ANF* 4:192).

10. Lactantius, *The Divine Institutes* (*ANF* 7:186–87).

11. Cyprian, "Cyprian's Answer to Cornelius, Concerning the Crimes of Novatus" (*ANF* 5:325–26).

12. "Womb binding" was used to induce abortion particularly during the third century. A binding cloth was tightly wrapped around the abdomen in order to prematurely expel the baby from the womb. The early church father Origen also spoke on this practice with derision in his work *Against Heresies* 9 (see Jones, *Soul*, 37).

have a child either by a slave or by any paltry fellow, for the sake of their family and excessive wealth. Behold, into how great impiety that lawless one has proceeded, by inculcating adultery and murder at the same time!"[13]

- In the fourth century, the practice of abortion became endemic among Christians. Instead of condoning it or using an argument that distinguished between killing formed and unformed fetuses, Christian leaders condemned it. Early church father Jerome (ca. 342–420) condemned abortion regardless of the age of the unborn (22.13).[14] Ambrose (339–397), the bishop of Milan, joined Jerome in his condemnation of abortion.[15]
- John Chrysostom (347–407), the archbishop of Constantinople, blasted Christian men, married or not, for visiting prostitutes. In his *Homily on Romans* (24), Chrysostom stated that such men were guilty not only of adultery, but also of murder if the prostitute became pregnant and had an abortion: "For even if the daring deed be hers, yet the causing of it is thine." He labeled abortion as "something even worse than murder" because it defiled a blessing like childbearing and degraded it into the murder of one's own offspring—and turned the chamber of new life, the womb, which he termed "the chamber of procreation," into "a chamber for murder."[16]
- Several Christian teachers and bishops noted the harmful character of abortion. Clement of Alexandria (153–217), head of the Catechetical School in Alexandria, wrote that those who obtain an abortion destroy not only a child but also "destroy all humanity" (2.10).[17] Augustine (354–430), bishop of Hippo, regarded abortion not only as an attack on human life but also as an attack on marriage. If a married couple agreed to abort their child, he wrote, "they are not husband and wife" (XV).[18]

13. Hippolytus, *Refutation of All Heresies* (*ANF* 5:129–31).

14. Jerome, "Letter XXII To Eustochium" (*ANF* 2:27).

15. Ambrose, *Hexameron*, 5.18.58.

16. John Chrysostom, *Homily on Romans*, in *Nicene and Post-Nicene Fathers*, 1st series (hereafter cited as *NPNF*), 11:520.

17. The Latin text reads, "*omnem humanitatem perdunt*." Clement of Alexandria, *The Instructor* (*ANF* 2:262).

18. Augustine, *On Marriage and Concupiscence* (*NPNF*[1] 5:271).

- Several Christian teachers even referred to abortion as *parricide*. Parricide typically refers to the intentional killing of a parent or relative, and their use of this word is rather strange, since abortion is the killing of an unborn child by a parent. However, parricide dredged up a particular horror in the Roman mindset, since it represented an attack upon authority. Minucius Felix, Cyprian, Lactantius, and Ambrose inverted this sense of the word and associated the horrors of parricide with an attack upon the weak and powerless. This reflects the Christian mindset with respect to abortion.

The early church clearly regarded abortion as a heinous crime. However, Jesus' death on the cross provided the means of forgiveness for those who had sinned, including those who had procured an abortion.

What did the church recommend for someone who had had an abortion? The fourth century provides the first evidence of a codified attempt to restore those who had had an abortion. The Council of Elvira in Spain (ca. AD 305) issued eighty-one canons, two of which dealt with abortion. Canon 63 held that a woman who has an abortion should be kept from communion "even as death approaches, because she has sinned twice"—once for adultery and again for muder. In Canon 68, any Christian who had not yet been baptized and became pregnant and had an abortion "may be baptized only when death approaches."[19] The canons on abortion were quite severe, but no more severe than canons on other subjects accepted at the synod.

After AD 313, Emperor Constantine declared a policy of religious toleration. In AD 314, a body of bishops met at the Council of Ancyra in the Roman province of Galatia. One canon mentioned abortion.

> Canon 21: Concerning women who commit fornication, and destroy that which they have conceived, or who are employed in making drugs for abortion, a former decree excluded them until the hour of death, and to this some have assented. Nevertheless, being desirous to use somewhat greater lenity, we have ordained that they fulfill ten years [of penance], according to the prescribed degrees.[20]

19. Dale, *Synod of Elvira*, 129–32, 334–36.

20. Council of Ancyra, "Canon XXI," in *Nicene and Post-Nicene Fathers*, 2nd series (hereafter cited as *NPNF*^{2}), 14:73.

This council condemned abortion and also held responsible those who made the drugs used to induce the abortion. However, it also prescribed a more lenient penance for those who had procured an abortion.

In AD 375, Basil the Great (330–379) wrote a series of letters that significantly influenced church legislation. Of these canons of Basil the Great, two are concerned with abortion:

> Canon 2: Let her that procures abortion undergo ten years' penance, whether the embryo were perfectly formed, or not.

> Canon 8: . . . But the man, or woman, is a murderer that gives a *philtrum*,[21] if the man that takes it die upon it; so are they who take medicines to procure abortion; and so are they who kill on the highway, and rapparees.[22]

Basil asserted that those who had an abortion or administered the drugs to induce the abortion were as guilty of murder as those who waylaid travelers and killed them. While there was disagreement as to the duration of the penance for such an act, there was no question as to the act's criminality. According to Basil, abortion is murder and it should be treated as such. For the purpose of penance, Basil rejected the distinction between the unborn being "formed," or "unformed." This distinction was not discussed at either the Council of Elvira or the Council of Ancyra, but Basil seems to have heard of it and "reveals a certain impatience with it."[23]

These canons of Basil the Great represent the first time in the literature of the early church that a distinction between a formed or unformed unborn is mentioned. This distinction originated in the Greek translation of the Hebrew Old Testament, known as the Septuagint, and in particular Exodus 21:22–25. Note a translation of these verses from Hebrew into English:

> And when men strive, and have smitten a pregnant woman, and her children have come out, and there is no mischief, he is certainly fined, as the husband of the woman doth lay upon him, and he hath given through the judges; and if there is mischief, then thou hast given life for life, eye for eye, tooth for tooth, hand for

21. *Philtrum* in ancient Latin medicine was a love potion.

22. Basil, "The First Canonical Epistle" (*NPNF*² 14:604–5). The term *rapparees* refers to bandits or robbers who prey on travelers.

23. Connery, *Abortion*, 49.

> hand, foot for foot, burning for burning, wound for wound, stripe for stripe.[24]

Now consider an English translation of these verses after they were rendered into Greek by the Septuagint translators:

> Now if two men fight and strike a pregnant woman and her child comes forth not fully formed, he shall be punished with a fine. According as the husband of the woman might impose, he shall pay with judicial assessment. But if it is fully formed he shall pay life for life, eye for eye, tooth for tooth, hand for hand, foot for foot, burn for burn, wound for wound, stripe for stripe.[25]

There are some significant differences between these two translations. The Hebrew text makes a distinction between the death of the fetus and the death of the mother, but the Septuagint distinguishes between the formed and unformed fetus. The Greek text prescribes the death penalty for causing the death of the formed fetus and does not mention the death of the mother.

Why the change? There is a good chance that this translation is an interpretation of the Hebrew text through the eyes of the philosophy of Aristotle.[26] Aristotle argued that the fetus became a full human being only when it was formed because that was when the human soul was infused into it.[27] He claimed that this occurred forty days after conception for a male fetus and ninety days after conception for a female fetus.[28]

Because the Septuagint was the Scripture for Greek-speaking Jews, this view of the unborn found its way into the works of several prominent Jewish documents. For example, the first-century Jewish philosopher Philo of Alexandria demarcated formed from unformed fetuses,[29] and some of the Talmudic literature held to views of the unborn that greatly resembled those of Aristotle.[30] Please note that this is not the only view of the unborn among ancient Jews. The Jewish literature shows some diversity of views,

24. Young, *Young's Literal Translation of the Bible*, Exod 21:22–25.

25. Pietersma and Wright, *New English Translation of the Septuagint*, Exod 21:22–25.

26. Connery, *Abortion*, 17–18.

27. Aristotle, *On the Generation of Animals*, II, 3.

28. Aristotle, *On the History of Animals*, VII, 3.

29. Philo of Alexandria, *On Special Laws*, III, 108–9.

30. *Babylonian Talmud Niddah* 15b–16a. Also see Feldman, *Marital Relations*, 266, n. 81.

but none of the ancient Jewish communities seemed to support abortion on demand except in cases where the mother's life was in danger.[31]

After Basil the Great wrote his highly influential canons, other councils produced results that contributed to the debate over the unborn:

- The Spanish Council of Lerida in AD 524 condemned adulterers (male and female) who conceived children out of wedlock and then tried to do away with them by abortion, and the *venefici*, or those who gave the poisons to produce an abortion. The penance for the adulterers was seven years, but for the poisoners it lasted a lifetime.[32]
- The Council of Trullo, which met in Constantinople in AD 692, was influenced by the canons of Basil the Great. Canon 91 of this council reads: "Those who give drugs for procuring abortion, and those who receive poisons to kill the fetus, are subjected to the penalty of murder."[33]

Because many of these canons on abortion were the result of local councils, collections of the conclusions of these councils appeared in the middle of the fifth century. Three councils in particular—Elvira, Ancyra, and Trullo—were included in most of these collections and commanded great respect for several centuries. In fact, these canons continue to inform canon law in the Orthodox Church to the present day. At no time in its ethical or legal history has the Eastern Orthodox Church embraced an ethical or legal distinction between early and late abortion.[34]

In conclusion, the earliest Christian thinking on this issue made no distinction between formed and unformed fetuses, but condemned abortion at all stages of development. The division between early unformed and later formed babies did not play any significant role in the first several centuries of Christian thinking, and was not even mentioned until the fourth and fifth centuries. This shows us why Dunstan's thesis is simply mistaken.

I hope this does the job.

Dr. B.

31. Jones, *Soul*, 43–56.
32. Bingham, *Sixteenth Book of the Antiquities of the Christian Faith*, 390–91.
33. Council of Trullo, "Canon XCI" (*NPNF*² 14:404).
34. Jones, *Soul*, 65.

Letter #4a

What about the Medieval Church?

Dr. B.,

I'm sorry but that answer did not do it. I am convinced now that the church's first take on the unborn was to value them from the earliest stages, but what about later? Is this book my mother loaned me right? Did the medieval church buy the formed/unformed distinction, and if so, when did they change their mind and why?

Another thing I don't get is this: If the later church changed its mind on this issue, then wasn't the later switch by Pope Pius just a new trend that this particular pope started?

Sorry to keep bugging you with this.

Rachel K.

Dear Rachel,

You should never be sorry to "keep bugging" me with such a subject, because this topic is really important. Bug me all you want on this one.

How and when did the switch occur? Those two questions are related. The answer is that the formed/unformed distinction took hold gradually. However, it is extremely important to remember that even during the Middle Ages, the more ancient tradition that regarded abortion as homicide regardless of the age of the unborn was still present and even enforced in some cases. The coexistence of these two views caused some confusion.

Some early church fathers mentioned the distinction between killing a formed and unformed fetus, but these statements were mixed with other declarations that supported the earlier tradition that made no such distinction. For example:

- Jerome shows a unified opposition to abortion. In his letter to Eustochium, he complains that some widows try to hide their failure to keep their vows of chastity by destroying human life before it has been conceived. He accuses a woman, who chemically sterilized herself, of murder. Others, Jerome says, try to abort the unborn baby and die in the process, thus committing three crimes: the violation of chastity, the murder of the child, and taking their own lives.[1] But in another letter to Algasius (Letter 121.4), he implies that only abortion of the formed fetus is murder.[2] What do we make of these two disparate letters?
- Likewise, Augustine in his *On Marriage and Concupiscence* classifies married couples who have sexual intercourse but try to prevent having children with adulterers and fornicators. If couples use sterilizing drugs and abortion-inducing concoctions, then they are not really married, but the woman is merely the man's mistress (1.17).[3] In his attack on Julian the Pelagian, Augustine asserts that never is the unborn baby considered a part of the mother because the baptizing of a pregnant mother does not cover her infant child, who must be baptized sometime after he is born.[4] In the *City of God*, he takes up the question

1. Jerome, "Letter XXII to Eustochium," in *Nicene and Post-Nicene Fathers*, ed. Philip Schaff, 2–6:27.

2. Connery, *Abortion*, 53; Jones, *Soul*, 65.

3. Augustine, "On Marriage and Concupiscence," in *Nicene and Post-Nicene Fathers*, ed. Philip Schaff, 1–5:270–71.

4. Connery, *Abortion*, 56.

> of the resurrection of aborted fetuses and sees no reason why they should be excluded (22.13).[5] However, in the *Enchiridion*, Augustine hedges on this same question and distinguishes between a formed and unformed fetus, ultimately concluding that whatever is missing from the unformed fetus will be supplied by God (85).[6] In his commentary on Exodus (80), he goes even further and suggests that the abortion of an unformed fetus is not homicide.[7] Augustine vacillated on this issue and never really came to a secure conclusion.

These disparate statements by these church fathers would provide future support for positions on the unborn that deviated from the original stand of the early church.

The differing views of the unborn first made their presence known in church discipline. In Mediterranean churches, the practice of penance was public. Penitents were excluded from taking communion or were excommunicated from the community of faith for a period of time. However, Christian communities in other corners of the world, particularly Ireland and England, practiced penance differently. In these places penance followed the model of the monks who confessed to their abbots. Penitents were never outside the community of faith. They practiced private rather than public penance. The development of private penance gave birth to *penitential literature*. Penitential books aided priests in assigning penances for different sins. These penitential books became quite common throughout the Western church near the close of the first millennium.[8]

Penitential literature also reflected changing perspectives on the unborn. For example, one of the earliest Irish penitential books, the Finnian, dates from the first half of the sixth century. The Finnian connects abortion with homicide and imposes a penance of six months on only bread and water, and a two-year fast from wine and meat, plus six forty-day fasts on bread and water for the person who caused the abortion. There was no penitential variation according to the age of the embryo.[9] However, in a later tradition from the late seventh-century Irish canons, other penitential

5. Augustine, "City of God," in *Nicene and Post-Nicene Fathers*, ed. Philip Schaff, 1–2:494.

6. Augustine, "Enchiridion," in *Nicene and Post-Nicene Fathers*, ed. Philip Schaff, 1–3:265.

7. Connery, *Abortion*, 57.

8. Ibid., 66–67.

9. Ibid., 70–71.

books, the Bigotian Penitential, and the Old Irish Penitential of the early ninth century prescribed penances for abortion that varied according to the age of the embryo or fetus when she was aborted. The Old Irish Penitential gave three and a half years penance for abortion after the pregnancy had become established, but seven years if the flesh had formed and fourteen years if the soul had entered the unborn child.[10] This three-stage process of development regarded the embryo as unformed "like water" in the first stage, with flesh but without a soul in the second stage, and ensouled and formed in the third stage. This threefold pattern is not found in the Septuagint or Aristotle, but in Hippocrates.[11] Thus for these writers, abortion was not true homicide if committed before the signs of movement or *quickening*, but was still regarded as a serious sin.

In the later seventh century, the first Anglo-Saxon penitential books were written. These made a distinction in penance if the embryo was older or younger than forty days old. The penance was more severe if the embryo was killed after forty days of life (three years, the same as a homicide), and less severe if the embryo was less than forty days old when destroyed (only one year of penance).[12]

These Celtic and Anglo-Saxon penitential books came into conflict with the older models of penance used in Spain. In AD 589, the Council of Toledo condemned the new penitential books as too lenient. Nevertheless, the penitential models used by the Anglo-Saxon and Celtic congregations gained ascendancy in the centuries to come, and by the time of the eleventh century, there were three models for treating abortion: 1) abortion was homicide regardless of the stage of the pregnancy; 2) abortion was homicide only after ensoulment; 3) both contraception and abortion were homicide regardless of the stage of the pregnancy.

The first option represents the dominant tradition from the *Didache* to the Council of Trullo, which was reaffirmed in the West in AD 848 in Canon 35 of the Council of Worms. The second and third options became increasingly influential from the late fourth century on. Unfortunately, thanks to the ambiguity of particular early church fathers, the same authors could be used to support multiple options. For example, Jerome could be used to support either option one, two, or three, depending whether you were reading Letter 22 or 121.

10. Ibid., 71–73.

11. Jones, *Soul*, 67–68.

12. Connery, *Abortion*, 72–74.

Collections of papal decrees called *decretums* also helped codify the formed/unformed distinction. Ivo of Chartes (1040–1115) cited Jerome and Augustine in his *Decretum* to support the notion that abortion before ensoulment of the unborn was not homicide. These same texts were used in 1140 by Gratian (d. 1159), and Peter Lombard (ca. 1100–1160). Gratian wrote *Concordance of Discordant Canons*, a collection that contained some four thousand items and was used as an authoritative source of canon law (the rules and regulations of the church). Gratian's work and Peter Lombard's *Four Books of Sentences* formed the foundation of church law and theology for the entire Middle Ages. In the thirteenth century, the Dominican Raymond of Pennafort (1175–1275) wrote a new canon law for Gregory IX, called the *Decretals*. He used the works of Ivo of Chartes and Gratian and followed the *Decretum* of Burchard of Worms with regard to contraception and abortion (*Decretals* V, tit. 12, can. 5).[13] Raymond also included a decision by Pope Innocent III with regard to a monk who, as a joke, had seized a pregnant woman's girdle and unintentionally caused her to miscarry (*Decretals* V, tit. 12, can. 20).[14] The pope decreed that the monk should be suspended only if the child was living (Latin: *vivificatus*). This produced a contradictory situation in which particular canons said one thing but a ruling from the same collection of canon law said a completely different thing.

To provide an example of the confusion that ensued from the coexistence of these two standards, Magister Rufinus (d. 1190) claimed that killing an embryo before ensoulment carried the guilt of homicide but not the act of homicide. Conversely, Roland Bandinelli (d. 1181) asserted that abortion was homicide regardless of the age of the pregnancy because the intent was the same in both cases, a position adopted by the Franciscan theologian Bonaventure (d. 1274). Thomas Aquinas (1225–1274) held that contraception and early abortion were second only to homicide. Aquinas wrote in *Summa Contra Gentiles*, "after the sin of murder, this sort of sin seems to hold the second place, whereby the generation of human nature is precluded."[15] Aquinas taught similar things in his *Commentary on the Sentences* (IV, D. 31 & Q.4).

Given the tremendous influence of his views, it is worth examining the thinking of Thomas Aquinas in greater detail. Thomas followed Aristotle's

13. Gregory IX, *Quinque Libri Decretalium.*

14. Ibid.

15. Thomas Aquinas, *Summa Contra Gentiles* III, Q.122.

account of the formation of the soul, and therefore thought that the embryo goes through a succession of souls; first it went through the vegetative, then the animal, and, finally, the human soul. The human soul is conferred at the culmination of development: "We conclude therefore that the intellectual soul is created at the end of human generation, and this soul is at the same time sensitive and nutritive, the pre-existing forms being corrupted."[16] Thus for Aquinas the embryo is truly alive but it does not possess human life until the human body is formed. In following Aristotle, Aquinas also thought that the embryo does not develop by means of a program that is directed from within, but from without. According to Aquinas, the power of development came from the father: "The formation of the body is caused by the generative power; not of that which is generated but of the father generating from seed, in which the formative power derived from the father's soul has its operation."[17] When the formation of that human body is complete, at forty days for a male embryo and ninety days for a female embryo, God creates the soul in the individual.[18] Aquinas' delayed view of ensoulment was completely dependent on Aristotelian science, which was completely rejected by the Reformer John Calvin, and repudiated by later embryological studies.

Abortion at any stage was condemned and the imposition of excommunication for abortion was upheld by local synods in Riez in 1234, Lille in 1288, Avignon in 1326, and Lavaur in 1368, but the prescribed penances continued to reflect the two-tiered system.[19]

In 1588, in a decree called *Effraennatum*, Pope Sixtus V called upon the power of excommunication in an attempt to restrain the growing practice of abortion during the Renaissance. He included contraception under this ban, since he used the *Decretals*, V.12.5 as his model. Three years later, Pope Gregory lifted the scope of this excommunication in his *Sedes Apostolicae*. He declared that only abortion of a formed fetus merited excommunication.

In 1869 Pope Pius IX issued a decree that protected the embryo from the moment of its creation until its natural death, and in doing so ended the confusion. However, this was not a novel decree, but the restoration

16. Thomas Aquinas, *Summa Theologica*, Ia, 118 art. 2 ad 2.

17. Ibid., IIIa, Q33 art. 1 ad 1.

18. Thomas Aquinas, *Commentary on the Sentences*, III, D3, Q.5 art.2. Aquinas cited Aristotle's *History of Animals* 7.3, 583b 3–5, 15–23.

19. Huser, *Canon Law*, 58–59.

of an ancient understanding with regard to abortion and the nature of the unborn. The Christian church was unique in its defense of the unborn throughout their various stages of development, and Pope Pius IX merely returned to the original intent and understanding of the early church.

The Reformers held different views of *ensoulment*—when a human embryo possesses a soul—but they were rather unified in their view that the unborn are human persons worthy of protection from the beginning of their lives. In doing so, they held to the ancient rather than the derived view.[20]

That is probably a far longer answer than you wanted, but it was necessary to make the point. Feel free to "pester" me again if you need further clarification.

Michael Buratovich

20. Jones, *Soul*, 141–55.

Letter #5

The Smallest of Us

Dear Dr. Buratovich,

I am struggling to think through this and you said that we could ask, so I'm asking.

I read through all my notes as I got ready for Friday's quiz on fertilization. You said over and over in class that when fertilization ends, a new human being has come into existence and this new human being deserves our protection. If I heard you right, killing it would be the moral equivalent of murder.

Well . . . why? It is so small, so different from me, so puny and insignificant in its appearance that it seems far too different from me to be like me. How could killing this small cell be the same as killing the person who rides next to me on the bus? I'm having a genuinely hard time figuring this one out.

Answer me after the quiz on Friday because I probably won't read your e-mail until after the quiz.

Thanks,

Anna C.

Dear Anna,

Yes, you definitely heard me right, and thank you for listening so well. I gave a brief defense of my statement in class, but since I have *so* much to cover in lecture, it was just that—brief. A question as important as this certainly needs a more involved defense, and that is why I am answering so quickly even though you won't read it until after Friday.

Let's first start with a short thought experiment on embryos. When I showed those pictures of sea urchin embryos in the laboratory, they didn't look like sea urchins, did they? The same goes for the fruit fly and chicken embryos I showed you and your colleagues. They didn't look like adult fruit flies or chickens. However, what are they? That is to ask, "What kind of thing are they?" If you were to think about it a little bit, I hope you would see that a sea urchin embryo is not a different kind of thing from an adult or larval sea urchin. It is still a sea urchin, albeit a very young sea urchin. The same goes for the fruit fly or the chicken. These fruit fly and chicken embryos are very young fruit fly and chicken embryos, and not a different kind of thing. A sea urchin embryo is not a carrot, or a zebra; it is the same type of thing as the adult of its species, just at an earlier stage in its life cycle. This same rule applies to human embryos. They are not a different kind of thing than an adult, adolescent, toddler, fetal, or embryonic (eight weeks old or younger) human; they are simply quite a bit younger.[1]

Let's take this a step further. The single-celled zygote is a stage we have all passed through. You and I were both in high school at one time, we were both middle schoolers at one time, we were both toddlers at one time, and we were both newborn babies. But we can extend this even further. You and I (and everyone else we know) were once twenty-two-week-old fetuses, were once fifteen-week-old fetuses, were once six-week-old embryos, and we were all once zygotes. Consider yourself at that stage, Anna: were you the same entity then that you are now? Sure, you are larger and can do more things, but everything that you are today was present in that zygote in a *radical form*. This is to say that the zygote was like a Polaroid snapshot of you. You are a rational, free agent[2] now and you were developing into one then.

You are too young to remember the old Polaroid cameras, but when you took the picture, a photograph rolled out and developed right before your eyes. You could then determine whether the picture was good or not

1. Koukl, *Precious Unborn Human Person*, 21–25.

2. Lee, *Abortion and Unborn Human Life*, 59.

and then you could take another picture. At the completion of fertilization, the *film* of your life rolled out and began to develop according to a programmed process. It was influenced by many events along the way, but this development has led to the beautiful young adult woman that you are today. Just as the Polaroid picture that was taken is the same picture before and after its development, so are you the same person who began her life at the completion of conception.[3]

With respect to your statements about how different the embryo looks from you, we can look at this a little more deeply too. Consider the embryo and you at your present age: in what ways do you differ? If you think about it, there are really only four ways in which you differ from your embryo-stage self. You differ in size, level of development, environment (or location), and degree of dependence.[4] It really comes down to those four things. Now, ask yourself this question: "Do these differences make an embryo a different kind of thing so that we can justifiably destroy it?" Well, let's see.

What about size? When I lived in England I attended the same church as Julie Peters, the little person who starred in the movie *Willow*. She could not have been more than four feet, five inches tall (her sister was quite tall—go figure). Is Julie less of a person or of less value than the professional basketball player Shaquille O'Neil, who is seven feet tall? Certainly not. Shaquille O'Neil makes more money, but he is not more of a person because he is taller. Do children achieve more of a right to life as they grow taller? Are children with stunted growth syndromes inferior to those who grow? The answer to both of these questions is a resounding no. Therefore size is not a good indicator of a right to life.

Level of development is one that is used by lots of people, but this fails too. Can we kill infants, but not five-year-olds? Are ten-year-olds more valuable than eight-year-olds? You reach your physical prime in your twenties. Are twenty-year-olds more valuable than forty-year-olds? You reach your mental prime in your forties. So forty-year-olds are more worthy of life than sixty-year-olds, right? None of this works. Level of development is an equally unworkable criterion for granting someone the right to life.

Environment is even worse. Are kids in South Carolina as deserving of life as those in Alaska? What about those who live in huts in Burundi versus those in Kenya? Do kids change in value when their families move them? Are kids in cold climates more valuable than those in warm climates?

3. Klusendorf, *Case for Life*, 45–46.

4. Swarz, *Moral Question of Abortion*, 15–19.

Nonsense. Environment is an even worse criterion to determine whether someone has a right to life.

Degree of dependence? This doesn't work either. We start out utterly dependent on our parents and then become gradually more and more independent until we strike out on our own. We live as independent folks until the ravages of age rob us of our independence. Are these seasoned citizens less valuable than those who have yet to experience the inexorable consequences of age? No way. Do kids become more valuable as they become more independent? No way. This one doesn't work either.

Therefore, I think you can see that the main differences between the embryo and the young adult do not produce any difference in moral standing. The embryo is a human being at its earliest stages of life. How is it that scientists can say that sea urchin embryos are young sea urchins and fruit fly embryos are young fruit flies, but when it comes to humans, they are something so different that we cannot grant them the right to life? This seems to me to be a remarkable double standard.

Let me say one more thing about the human embryo. In class we discussed all the modifications to the genomes of the egg and sperm that cause the expression of the genes from the father's sperm to be expressed largely in the developing placenta and the genes from the mother's egg to be expressed largely in the developing embryo. This event, known as genetic imprinting, means that if an embryo possesses a genome that results from the combination of two sperm nuclei or two egg nuclei it will not develop normally. Human zygotes result from fertilization of an egg by a sperm, and at the culmination of fertilization they possess twenty-three pairs of chromosomes, half of which came from the mother and half of which came from the father.[5] Now sometimes fertilization does not occur properly, and in the majority of these cases, the result is something that possesses an abnormal number of chromosomes and that dies soon thereafter. However, occasionally, embryos with abnormal numbers of chromosomes do survive.

My music minister and his wife are the proud parents of a Down syndrome child. Is she a human person? Absolutely. Such individuals still possess the characteristics of humanity in a radical form, even if their ability to develop into a mature human being is somewhat handicapped. They can survive, develop, and live, and should be afforded full human status in spite of their handicap.

5. For a good explanation of genetic imprinting and its role in human development, see Schoenwolf et al., *Larsen's Human Embryology*, 60–67.

All other types of arguments that the embryo does not have desires, sentience, or interests and therefore cannot be a person ultimately fail because people who are 1) emotionally unbalanced, 2) temporarily unconscious, or 3) conditioned to not desire things are still persons even though they lack desires.[6] Other arguments that they lack consciousness also fail because the embryo will have consciousness, and therefore possesses it in a Polaroid form. Tell me, can you think of anything that can develop full-blown human consciousness and yet is not a human person?

The human embryo is a full member of the human community; it is just immature. Immaturity does not disqualify people from personhood outside the womb. Why does it disqualify humans from personhood inside the womb? Now we are back to an argument from environment, which does not work.

This is my defense of the embryo as a human person. I do not know if it convinces you, but I think that it is at least reasonable.

Thanks for asking. I'll see you Friday. Don't forget the quiz!

Michael Buratovich

6. Lee, *Abortion and Unborn Human Life*, 11–23.

Letter #6

But Too Many of Them Die!

Hey Prof,

I have read this over and over again: 50 to 80 percent of all embryos die before they ever implant into the uterus. I think this torpedoes your "embryo is a human person" shtick. If the mother's body, by nature, gets rid of well over half of the embryos, where are God's intentions in the killing of so many unique humans? It makes more sense to me that the early embryo is not an individual human person, but only becomes one later.

How do you like them apples?

James G.

Dear James,

This is a really commonly made argument. For example:

> The mother's body does not necessarily honor the moment in which a unique genome is established with quite the respect the ethicists do. Estimates range from 50 percent to 80 percent of naturally fertilized eggs are flushed from the mother's body before they can adhere to the uterine wall. Ponder just how many unique genomes get flushed right out of the system! If the Vatican is serious about associating a divine soul with each and every zygote, and if the mother's body by nature eliminates the majority of ensouled embryos, then theologically it would be difficult to see intentions as carried out by natural processes. An estimated 50 to 80 percent of all embryos die before they implant into the uterine wall.[1]

This extensive loss of life at this early embryonic phase suggests to some that there is no individual human person at this stage.[2]

I agree that this argument has intuitive appeal, but it has lots of problems.

First of all, the percentage of *normal* embryos that die before implantation is almost certainly inflated. Many embryos that die early on are grossly abnormal. The process called fertilization includes the fusion of the egg and the sperm. This step is necessary to form the early one-celled embryo (known as a zygote), but it is not sufficient. Fertilization includes several other steps as well.[3] The onset of the first cell division (cleavage) signals the completion of fertilization (traditionally known as "conception"). Once conception is completed, the embryo usually contains twenty-three chromosome pairs, half of which were contributed by the sperm and the other half of which were provided by the egg. The exceptions to this are those embryos that can still survive and develop, like in the case of embryos with an extra copy of chromosome 21. Abnormal conception creates abnormal embryos, some of which possess abnormal numbers of chromosomes. Scientists refer to cells that contain too many or too few chromosomes as "aneuploid."[4] Aneuploid embryos usually die soon after their inception. Since human life starts at conception, embryos that have not gone through a normal conception are not human persons, with the exception of those

1. Peters, *Stem Cell Debate*, 40–41.
2. Shannon and Wolter, "Moral Status of the Pre-Embryo," 619.
3. Kiessling and Anderson, *Human Embryonic Stem Cells*, 50–67.
4. Hartwell et al., *Genetics*, 516–18.

embryos with abnormal chromosomal numbers that still allow them to survive and continue development.

Does aneuploidy affect human embryos on a regular basis? Unfortunately, the answer is yes. Studies of embryos generated by in vitro fertilization have established that embryos that develop quickly implant relatively efficiently (approximately 50 percent), while those that develop slowly implant rather poorly (approximately 10 percent).[5] Slow-growing embryos often suffer from aneuploidy. Prescreening of embryos formed by in vitro fertilization has shown that up to 50 percent of all embryos are aneuploid.[6] Remember that reproductive specialists screen embryos during in vitro fertilization by looking at them under a microscope and grading them according to their appearance. However, these embryos with abnormal chromosome numbers often appear completely normal under the microscope. Therefore, you cannot assess the quality of embryos by simply looking at them.[7] This shows that the estimated percentage of normal embryos that die before implantation is certainly inflated.

Even if we grant this percentage, does a high mortality rate indicate that the early embryo is not a human person? Consider that babies from poorer countries experience much higher mortality rates than babies from countries in the developed world. Does this mean that third-world babies are less human than their counterparts in the developed world? Surely not. Secondly, the mortality of all people is 100 percent no matter when they die. Does this affect our assessment of the humanity of people in general?[8] Certainly not. Third, this argument assumes that early embryos are not valuable in the eyes of God because so many of them die, which defies the early church's understanding of Jesus' words in Matthew 25:40 to show concern for the "least of these." The Gospel shaped the Christian concern for society's weakest and most vulnerable individuals, and this concern was fundamental to their ethics. Early Christians consistently saved abandoned or deformed infants and strove to protect those still hidden in the womb. If early embryos are subject to high mortality rates, then the proper Christian response should be to deeply value them as some of the most vulnerable members of the human community.

5. Edwards et al., "Alleviating Human Infertility," 752–68.
6. Márquez et al., "Chromosome Abnormalities," 17–27.
7. Munné, "Chromosome Abnormalities and Human Embryos," 234–53.
8. Beckwith, *Defending Life*, 75–77.

In conclusion, high mortality rates are not morally significant, and do not tell us if the embryo is an individual human person.

That's how I see it. Be sure you let me know how your graduate school interview goes. I wrote you a sterling letter of recommendation. They have to accept you now.

Michael Buratovich

Letter #7

Twinning, Embryo Fusion, and Personhood

Dear Dr. Buratovich,

This has been bugging me for some time. Can you help me with this one, please?

Human embryos have abilities that argue against their being individual human beings. For example, the early embryo has the ability to split into twins and form two organisms where there was once only one. Also, two distinct embryos can fuse together to form one embryo that is a mixture of cells from two different eggs and two different sperm. How can entities that can split into two or fuse into one be regarded as human individuals?

Brooke C.

Dear Brooke,

Good question! Twinning is probably one of the most popular objections to the humanity of the embryo, but I think that it is one of the weaker arguments.

Consider this: how many twins were in your graduating high school class? No matter what your answer might be, it will represent a small percentage of the total people in your graduating class. Twins occur in approximately one in eighty live births, which means that about one in forty babies is born as a twin. The vast majority of these twins are "dizygotic" twins, which arise from two distinct eggs that are fertilized by different sperms and develop concurrently. "Monozygotic" twins result from the splitting of an early human embryo into two distinct but genetically identical embryos that develop in parallel. Monozygotic twins account for about one in 350 live births (approximately 0.3–0.4 percent).[1] This illustrates the largest problem with the twinning argument: Only a small proportion of human embryos show the ability to twin. Can we use such a rare event to make a blanket statement about the nonhuman status of all human embryos?[2] One in 150 children has autism, but it would be completely illegitimate to take the symptoms of autism and argue that this is the way young children in general act. Such a conclusion is not valid for either autism or twinning.

There are additional concerns. First, monozygotic twinning runs in families and there seems to be a strong genetic component to it.[3] Physician and theologian Edwin Hui interprets the inherited tendency of a minority of embryos to twin in this way: "twins are genetically determined and hence the two beings that emerge as twins are in actuality two from conception, although in 'latent' form."[4] This interpretation is certainly reasonable, and it means that terminating these embryos would end two human lives and not just one.

Twinning rates also increase when the embryo is physically manipulated outside the mother's body. Babies conceived through in vitro fertilization show two to five times the rates of monozygotic twinning.[5] In this

1. Hall, "Twinning," 735–43.

2. Hui, *Beginning of Life,* 69.

3. Machin, "Familial Monozygotic Twinning," 152–54.

4. Hui, *Beginning of Life*, 70.

5. Geoffrey et al., "Identical Twins and In Vitro Fertilization," 114–17; Sills et al., "Assisted Reproductive Technologies and Monozygous Twins," 217–23; Aston et al., "Monozygotic Twinning Associated with Assisted Reproductive Technologies," 377–86.

case, twinning is a response to slight injuries. The embryo responds by repairing the damage, and in doing so, making a second individual without the benefit of sexual reproduction. Therefore, the ability of the embryo to form a twin does not indicate the absence of a human individual, but the remarkable ability of an organism to heal and restore those bits of itself that were damaged or lost.[6]

Finally, Westmont College professor emeritus of philosophy Robert Wennberg assesses the twinning argument with a simple thought experiment. What if humans were able to twin at a stage of life other than the embryonic stage? Would we use this as a reason to regard them as a human nonperson? What if, for example, teenagers were able to twin right up to their sixteenth birthday? Would we think that they were not individual persons until they lost this ability? Clearly we would not because teenagers are the same people after their sixteenth birthdays as they were before. Similarly, if toddlers could twin until they finished the "terrible twos," would we regard them as human beings but not persons until they lost the ability to twin? Certainly not, for the same reason: toddlers who finally stop throwing tantrums are the same people they were before they became a little wiser. In both cases, the teenagers and toddlers would simply lack the ability to make a twin.[7] Since the lack of the ability to form a twin would not disqualify them from being a person, there must be something else that makes them a person.

Remember, Brooke, we lose abilities as we age, but that does not make us any less a human person. When your grandfather started using a walking cane, he did not become any less of a person. Likewise, the loss of the ability to form a twin is of no moral consequence for the status of the embryo.

While it does not receive the press that the twinning argument gets, embryo recombination is also used to argue against the personhood of human embryos. Two early human embryos can sometimes fuse to form one embryo composed of a mixture of two types of cells with distinct genetic fingerprints. This phenomenon is referred to as *chimerism*, and there are approximately thirty confirmed cases of it. While there are almost certainly more people with chimerism in the general population, the low number of cases indicates the rarity of this phenomenon. Chimeric people usually have two types of red blood cells circulating through their bodies,[8] or

6. Hurlbut, "Altered Nuclear Transfer," 211–28.
7. Wennberg, *Life in the Balance*, 71.
8. Watkins et al., "Human Dispermic Chimaera," 113–28.

patchy skin or eye pigmentation.[9] If a female embryo fuses with a male embryo, the chimeric individual can have sexual organs that are ambiguous or male and female sexual organs in the same body (hermaphroditism).[10] In other cases the chimeric person is completely normal.[11]

Does chimerism mean that human embryos are not human persons? Wennberg extends his example for twinning to teenagers who can somehow fuse to form one individual. In such a case, we would not regard such teenagers as nonpersons until they lose this ability. To be consistent, we should not regard the embryo as a nonperson either just because a stark minority of them has the ability to fuse.[12]

What happened to the embryos that fused? They died and a new individual rose from their ashes. Given the rarity of chimerism, the embryo does not possess an active capacity to fuse, but undergoes fusion under rare conditions in order to repair itself. If it is invalid to impute the ability to twin to all embryos, then it is surely unreasonable to impute an even rarer event like chimerism to all embryos.

Even if the embryo underwent chimerism on a regular basis, should we classify it as a nonperson? Since you are a *Star Trek* fan, I will give you this example. An episode of *Star Trek: Voyager*, titled "Tuvix," provides an excellent perspective on this question. In this episode, two members of the Voyager crew, Tuvok and Neelix, return to the ship by transporter, only to have the malfunctioning device merge the two individuals into one new person named "Tuvix." The *Voyager* crew eventually accepts Tuvix, but some crew members devise a way to separate the two individuals to their original states. However, Tuvix does not wish to die. Both Neelix and Tuvok were individuals before this event and Tuvix is also a person. The fusion of the two crew members did not affect their personhood before fusion occurred, and it does not affect the personhood of human embryos either.

Thanks for your question. See you in class tomorrow.

Michael Buratovich

9. Sybert, "Hypomelanosis of Ito," 141S–43S.

10. Green et al., "Chimaerism," 816–17; Strain et al., "True Hermaphrodite Chimera," 166–69; Repas-Humpe et al., "Dispermic Chimerism," 431–34.

11. Yu et al., "Identification of Tetragametic Chimerism," 1545–52.

12. Wennberg, *Life in the Balance*, 71.

Letter #8

The Cells of the Embryo are Totipotent

Dear Dr. B.,

In class, last Friday, you told us that the cells of the early embryo (blastomeres) are totipotent, which (I hope I get this right) means that they can form any adult or embryonic tissue. If this is true, and I didn't make a mess of it, then isn't the early embryo actually a clump of separate cells, each of which can become another embryo? In other words, the embryo is really only a bunch of individual cells rather than a distinctive, whole, human person. How then can a newly fertilized embryo be a human person?

Chris R.

Dear Chris,

Great question.

Yes, you did get it right—the blastomeres of the early embryo are totipotent,[1] and this means that they can make any cell type in the embryo or adult organism. This fact has persuaded some scholars that the early human embryo is really a "cluster of distinct individual cells, each of which is a centrally organized living individual."[2] A problem with this argument is that during normal embryonic development each blastomere of the early human embryo does not routinely form a mature human organism.[3] They only do that when they are separated from each other. If individual blastomeres from four- or eight-cell stage embryos are isolated in the laboratory and cultured, they can form separate, distinct human embryos.[4] However, this is not their normal mode of development. During normal development, the blastomeres clearly are interacting with each other to prevent them from becoming separate organisms. Collectively, these blastomeres act as a whole organism that follows a self-directed developmental program that moves it toward an internally specified goal. Additionally, the early human embryo is surrounded by a thick, fibrous membrane called the zona pellucida, and this holds the blastomeres together and prevents them from detaching from each other.

There are other reasons to doubt this totipotency argument. Philosopher Patrick Lee gives the example of a flatworm, which, when cut in half, regenerates to form two distinct worms. These cells of the flatworm can de-differentiate to form all the tissues of the adult flatworm. In other words, the cells of the two halves of the flatworm are totipotent, but this "does not imply that prior to the division the flatworm is merely an aggregate of cells or tissues."[5] Likewise, just because the earliest blastomeres of the human embryo are totipotent does not mean that it is an aggregate rather than an organism. The earliest stages of human development require cells that are totipotent because the embryo must make everything from sperm to skin. As the embryo grows, the potency of the blastomeres decreases as cells become more and more committed to particular developmental fates. The fact that the embryo does certain things by certain times illustrates its goal-

1. Strelchenko and Verlinsky, "Embryonic Stem Cells from Morula," 93–108.
2. Ford, *When Did I Begin?* 139.
3. Ashley and Moraczewski, "Human Rights Present from Conception?" 33–60.
4. Van de Velde et al., "Four Blastomeres Develop into Blastocysts," 1742–47.
5. Lee, *Abortion and Unborn Human Life*, 93.

directed behavior, which is characteristic of a whole, integrated organism rather than an aggregate of separate individuals.

See you in lab on Tuesday,

Michael Buratovich

Letter #9

It's Just a Clump of Cells

Dr. Buratovich,

My brother and I have been going around and around on the embryonic stem cell issue for several weeks now. He asserts that the cells of the early embryo are all capable of becoming a whole new human embryo and therefore there is no way the human embryo can be a human person. He also adds that early embryos are a disorganized clump of cells that have nothing to do with each other. The cells of an individual communicate with each other, but the cells of the human embryo are just a clump of cells waiting to become a human person. They are not yet a human person. I don't know what to say. Is he right?

John H.

Dear John,

I'm glad to hear that you are talking about the stem cell issue with other people, like your brother. This is how we hone our arguments even though it is sometimes difficult. With respect to your brother's assertions, I think I can help.

Your brother is right when he says that the cells of the early embryo can make any cell type in the embryo or adult organism (totipotent). However, in class we talked about some of the reasons why this fact does not derail the personhood of the early embryo.

As to your brother's other argument, he seems to be on the same page as others who have asserted that the early embryo is not an integrated organism because there is no "causal interaction between the cells."[1] Essentially, the blastomeres seem to behave independently of each other, and therefore, the embryo is more like a bag of marbles than a whole organism.[2]

This claim ignores the goal-directed behaviors of the embryo. Princeton University professor of jurisprudence Robert P. George and University of South Carolina philosopher Christopher Tollefsen have described three goals of the early embryo: 1) get to the uterus and implant, 2) form the structures necessary for successful implantation, and 3) preserve its structure against the many hazards it might encounter. Even these three goals are subordinated to a larger goal, which is to get to a hospitable environment and grow and flourish.[3] The early embryo's actions are directed toward these goals. The embryo also prepares for future events. For example, at the two-cell stage, the blastomeres synthesize a cell adhesion protein called E-cadherin. E-cadherin acts like cellular superglue, and the two-cell stage embryo makes it in anticipation of compaction, which occurs two days later.

A second assumption of this argument is that all the blastomeres are the same, which is also incorrect. By the four-cell stage, each blastomere shows distinct differences in gene expression and protein localization. Protein localization differences begin in the egg, while it is still housed in the ovary. Human eggs, like the eggs of other animals, have a lower part of the egg, called the *vegetal pole*. The vegetal pole usually marks the bottom of the egg. The other side of the egg opposite the vegetal pole is known as the *animal pole*, which serves as the top of the egg. Several proteins are

1. Smith and Brogaard, "Sixteen Days," 55.
2. Van Inwagen, *Material Beings*, 153–54.
3. George and Tollefsen, *Embryo*, 151–52.

loaded in the animal pole of human eggs, and the position of these proteins does not change after fertilization and conception. The subsequent cleavages, however, cause inequitable distribution of these proteins, which has profound effects on the development of these blastomeres (fig. 9.1).[4] The blastomere deprived of these localized proteins will give rise to the trophoblast, the flattened, outer layer of cells that becomes the placenta.[5] The developmental fate of this blastomere is caused by the behavior of its predecessors. These distinct protein distribution patterns essentially direct one of these cells to form the trophoblast and the others to form the cells of the inner cell mass. In other words, these blastomeres are interacting with each other.

4. Antczak and Van Blerkom, "Oocyte Influences on Early Development," 1067–86; Antczak and Van Blerkom, "Fragmentation in Early Human Embryos," 429–47.

5. Hansis et al., "Marker Genes in Human Preimplantation Embryos," 577–83; Hansis and Edwards, "Initial Differentiation of Blastomeres," 206–18.

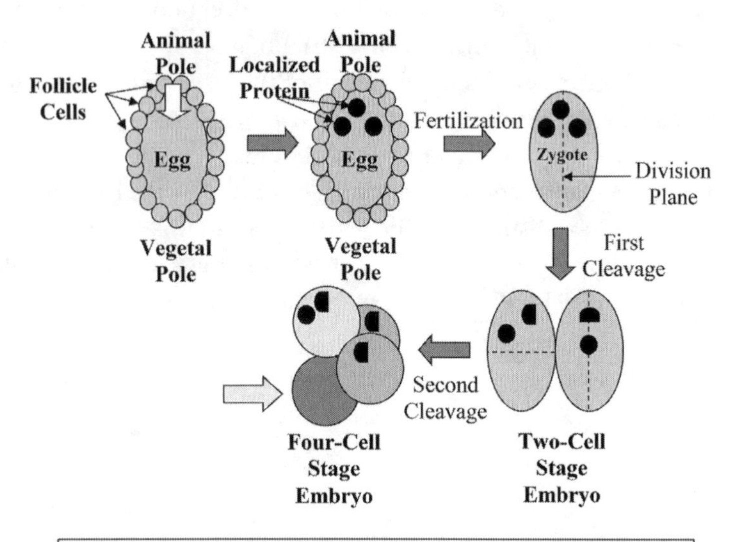

Figure 9.1 Protein localization in the early embryo is inherited from the egg. The egg is surrounded by follicle cells that transport proteins into the animal pole of the egg. These localized proteins remain in the animal pole after fertilization and conception. The first cleavage is meridional, which means that the division plane bisects the animal pole. This meridonial cell division equally distributes the localized proteins to each blastomere at the two-cell stage. However, the second cleavage is meridonial for one of the blastomeres and transverse, or parallel to the animal pole, for the other blastomere. This equally distributes the proteins into two of the blastomeres and unequally distributes them in the other two. The blastomere indicated by the arrow in the four-cell stage embryo will give rise to the trophoblast.

The upshot of all this is that the early embryo is not a clump of cells, but a dynamic, highly interactive, goal-directed entity. In this respect it sounds exactly like a human person, but is located inside the mother and is smaller, less mature, and more dependent than the rest of us—none of which are morally significant differences.

Please let me know how this affects your discussions with your brother.

Michael Buratovich

Letter #10

Fertilization Does Not Always Produce Human Entities

Doc B,

I saw my OB/GYN over the weekend and she gave me a clean bill of health. She also gave me something to think about. When she asked about school, I told her about your class and all the discussions we have been having about embryonic stem cells. She said that she supported researching embryonic stem cells because they can help a whole lot of people and the embryos that are sacrificed are going to die anyway. She also said she strongly doubted that early embryos become human persons at fertilization because fertilization sometimes produces something that is simply not human. She talked about "hydatiform moles" and "blighted ova" and stuff like that (I hope I spelled all that right). If fertilization really generates this kind of stuff, then how can we say that fertilization is the start of a human person? Doesn't something else have to happen to make the embryo a person then?

Brooke N.

Dear Brooke,

You have raised some excellent questions, and I am glad that you sought out a professional to get some answers. However, I must scold you slightly for not listening very well in class. These examples your OB/GYN gave were precisely some of the reasons I gave for insisting that fertilization could *not* be the stage at which a human person comes into being, *if* we define fertilization as the fusion of the egg and the sperm. However, if we see fertilization as the multistep process we discussed in class, then it is at the end of fertilization (or conception) that a new human person comes into existence. Let's review a little and extend the discussion we had in class.

Fertilization begins with the fusion of the egg and the sperm. Fertilization also encompasses a whole host of events that include egg activation, fusion of the egg and sperm nuclei, and preparation for the first cell division. Once conception is completed, the life of a new human person has begun.

Sometimes fertilization does not occur properly, and it leads not to a human person, but to something quite abnormal. The result is somewhat like a "hydatiform mole," "choriocarcinoma," or "blighted ovum." None of these resembles a normal human embryo. Since fertilization can create obviously nonhuman entities, it is unlikely that a human person is formed during the first stage of fertilization, but only at its completion. The fusion of the egg and a sperm initiates a multistep cascade, and abnormal or incomplete fertilization generates entities that are usually not human persons.

Normal conception usually produces entities with twenty-three pairs of chromosomes—one set that came from the mother and another that came from the father. Conception must also produce a zygote that is undamaged and can perform the subsequent stages of development. Sperm and egg nuclei contain chromosomes that undergo modification so that the chromosomes that come from the sperm are predominantly expressed in the placenta and the chromosomes that come from the egg are predominantly expressed in the embryo proper. This process is called imprinting, and it guarantees that human embryos cannot develop normally unless they receive one set of chromosomes from the mother and another set from the father.[1] Without two sets of chromosomes, one from each parent, abnormal conception is ensured and an abnormal nonhuman entity usually results.

1. Sha, "Mechanistic View of Genomic Imprinting," 197–216.

The zygote also possesses machinery for the ensuing cell divisions. If these features are inoperative, then the embryo has not really formed.

Hydatiform moles have no identifiable embryonic or fetal tissues and consist of grape-like sacs of blistered placental structures. There are complete and partial hydatiform moles. In the majority of complete hydatiform moles (approximately 60 percent), the egg pronucleus disintegrates or is inactivated. Fertilization by a single sperm is followed by duplication of the sperm pronucleus to generate a cell with two sperm pronuclei rather than an egg pronucleus and a sperm pronucleus. In other cases (approximately 20 percent), an egg with no nucleus is fertilized by two different spermatozoa. Egg cells with two sperm nuclei are known as "androgenotes," and they grow to form a hydatiform mole.[2] The remaining cases of hydatiform mole possess an egg and a sperm pronucleus, but are found in mothers who lack a functional copy of a gene called *NALP7*. Defects in this gene, if inherited, can cause inherited cases of complete hydatiform mole called "familial biparental hydatidiform mole." Without functional NALP7, the mother makes defective eggs that cannot support embryonic growth.[3] Even if fertilized, the egg undergoes abnormal conception and aberrant growth.[4] Partial hydatiform moles usually result from cells that have three copies of each chromosome—one from the mother and two from the father.[5]

Blighted ova have only placenta and no embryo, and are also the result of abnormal conception. They result from cells that contain pronuclei from both a sperm and an egg, but often have abnormal chromosome numbers.[6] Other blighted ova have smaller chromosome abnormalities that are only detectable with high-resolution techniques.[7]

Abnormal conception can also form aggressive cancers called choriocarcinomas that invade the uterine wall. Four percent of all complete hydatiform moles can become choriocarcinomas. Choriocarcinomas result from abnormal conception that generates entities that lack the genes that

2. Slim and Mehio, "Genetics of Hydatiform Moles," 25–34.

3. Hayward et al., "Genetic and Epigenetic Analysis of Recurrent Hydatidiform Mole," E629–39.

4. Deveault et al., "Proposed Mechanism for Mole Formation," 888–97.

5. Devriendt, "Hydatidiform Mole and Triploidy," 137–42.

6. Slim and Mehio, "Genetics of Hydatiform Moles," 27.

7. Azmanova et al., "Chromosomal Aberrations in Different Spontaneous Abortions," 127–31.

control cell growth and consequently divide uncontrollably.[8] However, evidence also exists that choriocarcinomas result from aberrant imprinting,[9] which again points to a defective conception event.

Thus abnormal conceptions usually do not produce human persons. That is the lesson we can take from these examples. I agree with your OB/GYN that the fusion of egg and sperm (the first stage of fertilization) does not automatically produce a person. However, conception, or the completion of fertilization, does, and that gives me pause when it comes to embryonic stem cell research that kills fully conceived human embryos.

I hope I answered your questions.

Sincerely yours,

Michael Buratovich

8. Asanoma, "*NECC1*," 15–25.

9. Arima et al., "Imprinting with Choriocarcinoma Development," 39–47; Wake et al., "IGF2 and H19 Imprinting in Choriocarcinoma Development," S1–8.

Letter #11

Embryo Skepticism

Dear Dr. Buratovich,

I was talking to my friend who goes to a large university in New York about the stem cell issue. He said that there is no way that life can begin at conception because there is disagreement among embryologists as to when a new organism comes into being. Some embryologists say that it occurs before the fusion of the sperm and egg nuclei, and others say that it occurs after this fusion. If these experts cannot agree, then how can we agree? Furthermore, he argued, if we cannot know when a human being comes into existence, then how can we argue that a human being comes into existence after conception?

He seems to have a point. What am I missing, if anything?

William C.

Letter #11

Dear William,

Yes, your friend does have a point. In fact, his argument is much like the one used by philosopher David Boonin in his book *A Defense of Abortion*.[1] Essentially Boonin raises two questions: 1) When do we know the embryo is an individual? and 2) When do the sperm and egg cease to exist?

These questions do not detract from the fact that the cell that is made at the end of fertilization, which is known as a zygote, is a whole organism whose parts work together to drive that whole organism to grow and mature. Just because we have some disagreement as to when a zygote precisely comes into existence does not change what a zygote is.

Philosopher Frank Beckwith evaluates Boonin's argument in this manner: "It seems to me that Boonin commits the fallacy of the beard: just because I cannot tell you when stubble ends and a beard begins, this does not mean that I cannot distinguish bearded faces from clean shaven ones."[2]

Beckwith also points out that pro-abortion advocates use various standards like consciousness, having desires, and so on as properties that separate a being that has a right to life from one that does not. However, just because we cannot precisely determine when those properties come into being does not cause pro-abortion advocates to question them.[3] Therefore this objection only questions when a zygote comes into existence; it does not nullify the humanity of the zygote.

Finally, this objection is confounded by the facts of human development. Fertilization of the egg initiates a cascade of events that culminates in the conversion of the egg into a zygote and the division of the zygote into two cells. Therefore, the objection that we can never truly know when fertilization or conception ends is simply untrue. Once the zygote divides into two cells, conception has certainly ended. Therefore this objection clearly falls apart upon further examination.

Thanks for your question,

Michael Buratovich

1. Boonin, *Defense of Abortion*, 37–40.
2. Beckwith, "Defending Abortion Philosophically," 177–203.
3. Ibid., 181.

Letter #12

Every Cell in Our Bodies Has the Ability to Be an Embryo

Dr. Buratovich,

Your crusade against embryonic stem cell research is, in my view, quite wrongheaded. You have argued that the embryo is a human person, but we can remove the nuclei from cells in our bodies and insert them into human eggs that have had their nuclei removed. This transforms that egg into a human embryo that has the same genetic makeup as the person who donated the cell that served as the source of the nucleus. This means that every cell in our body is a potential embryo. Yet we do not lament the loss of dead skin cells, or blanch when we sneeze. Neither do we get too upset when women have a period and lose lots of cells, or when couples have sex and effectively cast aside millions of spermatozoa. I could go on, but I think you get the point. If you are not willing to regard all our cells as human persons then you cannot consistently regard the embryo as a human person either.

Sincerely yours,

Emily W.

Dear Emily,

The argument you have raised against the humanity of the embryo is very similar to the one used by Ronald Bailey, editor of *Reason* magazine.

Bailey bases his argument on the following facts:

1. The vast majority of the cells in our body contain a nucleus that is filled with twenty-three pairs of chromosomes.
2. The chromosomes are made of a molecule called deoxyribonucleic acid or DNA.
3. DNA stores genetic information in separate packets called genes. When genes are accessed or turned on, specific proteins are made. In muscle cells, muscle-specific genes are turned on and the cell makes muscle-specific proteins. The complete complement of all human genes is called the human genome.
4. Almost every cell in our body contains a full human genome, with twenty-three pairs of chromosomes and all the genes found in modern humans.
5. Our cells are specialized for specific tasks, and this specialization occurs because cells turn particular genes on, and shut off many others.
6. When the nucleus from a body cell is placed into an egg, it is reprogrammed and many of the formerly inactive genes are now turned on.
7. Occasionally, the transfer of a nucleus from a body cell into an egg that has had its nucleus removed leads to the formation of an embryo.

Bailey interprets these facts to mean that any cell in our body can become an embryo. He writes, "Each skin cell, each neuron, each liver cell is *potentially* a person. All that's lacking is the will and the application of the appropriate technology."[1] Thus almost every cell in our bodies is a potential embryo; all that it needs is sufficient genetic reprogramming.

If an embryo is a human person, then so are the cells in our bodies. Bailey quotes the Director of the Oxford Uehiro Centre for Practical Ethics at the University of Oxford, Professor Julian Savulescu: "If all our cells could be persons, then we cannot appeal to the fact that an embryo could be a person to justify the special treatment we give it."[2] Since we simply do

1. Bailey, "Are Stem Cells Babies?"
2. Savulescu, "Should We Clone Human Beings?" 87–95.

not regard discarded body cells as persons, we cannot classify embryos as such either.

A variation on this argument is given by Lee Silver, professor of genetics at Princeton University. Silver refers to *tetraploid rescue* experiments in mice that use normal mouse embryonic stem cells and mouse embryos that have twice the number of chromosomes (a condition called tetraploidy). Tetraploid mouse embryos cannot form the inner cell mass of the embryo that forms the body of the fetus, but they can form the placenta. If normal mouse embryonic stem cells are inserted into the interior of tetraploid mouse embryos, the tetraploid cells will make the placenta, and the embryonic stem cells will form the inner cell mass. A functional mouse embryo results from this procedure that gives rise to a normal mouse pup. Silver interprets the experiments in this way: "Embryonic stem cells can develop into an actual person."[3] Embryonic stem cells, therefore, would become mature members of the human species if only they were to have the proper environment. Since we do not normally regard embryonic stem cells as persons, neither should we regard embryos as human persons.

I think that these arguments are severely flawed for two main reasons. First, cloning techniques are more than the "application of an appropriate technology." During a cloning procedure, the somatic cell nucleus works together with the enucleated egg to form an embryo. The somatic cell nucleus needs more than just the right environment. This procedure transforms it from the inside out, and changes the somatic cell nucleus from one kind of thing to another. The cloning experiment results in the end of the somatic cell nucleus and enucleated egg cell, and results in the creation of a totally new entity—an embryo—from the contributions of two defunct entities. The somatic cell is not analogous to an embryo but to the eggs and sperm that make embryos. In contrast, when an embryo grows, develops, and matures it does not become something other than a human being.

Secondly, the cells of our bodies, or somatic cells, are not human persons, since they are not distinct organisms, but are instead part of a larger, distinct organism. The human embryo is not a part of an organism but is a distinct organism that possesses an intrinsic, self-integrating ability to develop into more mature stages. Somatic cells possess no such ability.

3. Silver, "The Biotech Culture Clash," *Checkbiotech* (blog), July 18, 2006, http://greenbio.checkbiotech.org/news/biotech_culture_clash_embedded_religious_perspectives_east_and_west_create_distinct_r.

Lee Silver's variation doesn't fare much better. His conclusion that embryonic stem cells can "by themselves" become an embryo is simply not warranted by the details of the experiment he describes; the embryonic stem cells are placed into a tetraploid embryo, and the embryo works together with the stem cells to form a mouse pup. The tetraploid embryo directs and shapes the behavior of the implanted cells to transform them into something other than what they are. This is a change from one kind of thing (embryonic stem cell) to a different kind of thing (embryo). The union culminates in the expiration of the two original entities and the creation of a completely new entity. This is not a case of embryonic stem cells forming a mouse pup "by themselves" or "on their own." Instead, they require substantial manipulation to do so. As George and Tollefsen note, "this is a bit like saying that a pile of bricks can produce by itself a house—perhaps the house requires no other material for its construction, but it is not the case that the bricks organize themselves and direct the process by which they 'become' a house."[4]

Our own life histories confirm these conclusions. None of us began life as an egg or a sperm; rather, each of us began as an embryo. Likewise, cloned individuals would not begin as a skin cell or an egg, but as human embryos. Therefore the identification of somatic cells as embryos fails to convince.

Finally, let's look at this practically. We can identify embryos because they are undergoing this incredible process called development. Our somatic cells are not undergoing development and neither are embryonic stem cells. Thus this identification is absurd on its face.

I hope this gives you something to think about.

Michael Buratovich

4. George and Tollefsen, *Embryo*, 165.

Letter #13

But the Embryo Does Not Have a Brain!

Hey B man,

I keep going over your lectures on early human development and our discussion about death and something does not make sense to me. If a human being is dead when the brain ceases to function, and since the embryo does not start to make a nervous system until after primitive formation and the onset of gastrulation, then how is it that the embryo is a human person? We define death as a human body that no longer has a working brain, but you are arguing that the brainless embryo is a human person. If having a brain is central to defining if someone is alive, then why is the early human embryo a human person before it develops a functional brain?

Collin H.

LETTER #13

Dear Collin,

This is a great question, like so many of the other questions you ask in class. This argument you have presented is similar to an argument made by Michael Gazzaniga, professor of neurology at Dartmouth College. Since modern medicine uses brain death as a criterion for when physicians can harvest organs from someone's body, the presence of a functioning brain seems to be where we draw the line between a living and dead person. Since the embryo falls on the dead side of this line, it is not a human person.

This argument assumes that we classify brain-dead people as dead because we have an equation that goes like this: no functioning brain = not a human person. However, the equation is actually very different. Physicians regard a brain-dead person as dead even if other systems are still being artificially maintained because the brain acts as an integrating center for many of the organ systems of the human body. Breathing and the heartbeat are all controlled by clusters of nerve cells called "vital centers" in the brain stem. When the brain dies, these systems also cease to function on their own, which explains the old boxer's saying "Kill the head and the body dies." Therefore a brain-dead person has ceased being an integrated, self-directed organism and has become a corpse. The equation should read: brain dead = not an integrated, self-directed organism.

The embryo, however, is an integrated, self-directed organism that is in the process of becoming a more mature human person. We should not say that the early embryo does not have a brain, but that it does not have a brain *yet*. This is a crucial distinction because the embryo has everything it needs to grow a human brain, but the brain-dead individual lacks all capacity to do such a thing. While the brain-dead individual is truly dead and in the process of decay, the embryo is growing, developing, and very much alive. The early embryo is radically different from a corpse, and any argument that denies this is trying to force a definition on it that simply does not correspond to biological reality. The life of a corpse is over, but the life of the embryo lies ahead of it.

Secondly, when it comes to the embryo and the human brain, some people attach a great deal of importance to the presence of a brain. Personhood is simply not realized if there is no functioning human brain. However, can you name anything that can grow a human brain that is not human? Clearly the answer is no. If it can grow a human brain, then it is a human person. Therefore the "presence of a brain" argument is unsatisfactory as well.

I hope this does it.

Keep asking great questions,

Michael Buratovich

Letter #14

The Embryo Is Like an Acorn, Not an Oak Tree

Dear Dr. B.,

Thanks for coming to our basketball game. It means a lot to us when faculty come. Even though we lost, we felt as though we played hard and represented Christ faithfully.

After the game I was talking to a fellow player who asked me about embryonic stem cells. Apparently this has been bothering her for a while. She said that she heard some guy argue that the embryo is like an acorn and not like an oak tree. We value the oak tree for its grandeur and beauty, and we feel really bummed when a storm knocks it down. But the acorn—well, that's a different story because we don't care if it dies. In the same way, we value the death of a loved one who is grown up, but no one bats an eyelash if an embryo dies. Since the embryo is like an acorn and not an oak tree, we should not get all torn up about embryos dying in the lab. I had no idea how to respond to this. Any ideas?

Tristen S.

Dear Tristen,

I thought you played brilliantly, and even though you guys lost, you gave them a definite run for their money. I wouldn't be surprised if you take them down next time. Keep at it.

With respect to your stem cell question, your teammate has hit upon an argument used by a gentleman named Michael Sandel, who is a professor of political theory at Harvard University and was a member of President George W. Bush's President's Council on Bioethics.

Sandel states that even though every oak tree was at one time in its life an acorn, "it does not follow that acorns are oak trees." Why? Consider how we might regard the loss of a magnificent oak tree versus the loss of a single acorn. While the loss of the oak tree is received with distress and disappointment, the loss of the acorn hardly evokes a response. The same kind of disparate response is evoked by the loss of a mature human being as contrasted with the loss of a human embryo. Therefore, just as an acorn is a different kind of thing than an oak tree, a human embryo is a different kind of thing than a mature human person.[1]

This argument has a lot of things wrong with it. In the first place, it begins with a significant biological error. Acorns and oak trees are both oaks, that is, members of the genus *Quercus*. They are the same organism at different stages of development. Sandel acknowledges this at first, and then denies it based on features that the oak tree acquires later in its development.

Secondly, Robert George and Franciscan University of Steubenville philosopher Patrick Lee have pointed out that the characteristics Sandel uses to categorize oak trees and acorns into different groups—like grandeur, beauty, and so on—are "accidental" attributes. This is just a fancy way of describing a feature that is incidental to who you are and not essential. You can change an accidental attribute of something without affecting what it is. A good example is skin color in humans. People can have red, yellow, black, or white skin and still be people. If their skin color changes—for example, when they get a tan in the summer and lighten up in the winter—they are still people. Herein lies the reason why it was evil for people with lighter skin to enslave those who had darker skin. Even though some argued that people with dark skin were not human persons, this is plainly ridiculous because a person can have light or dark skin and still be a person; skin color is an accidental attribute. Thus, a scrawny, scruffy, diseased

1. Sandel, "Embryo Ethics," 207–9.

oak tree with twisted bark is just as much an oak tree as a large, grand one. The grandeur, size, and beauty of the tree are accidental features that do not make it an oak tree. Sandel has made accidental features of oak trees and humans the most important things about them even though they are not.

George and Lee make other telling points against Sandel's analogy. If acorns are to embryos as oak trees are to people, then what about oak saplings? Forest managers often cull oak saplings to prevent excessive crowding of trees and promote the health of the forest, and no one has any misgivings about such a practice. Yet if acorns are like human embryos, then oak saplings are like human toddlers. We would not entertain culling toddlers. We would also not have any trouble with pulling up and burning a diseased, disheveled oak tree, but killing the mentally or physically disabled would be just plain incorrigibly evil. Clearly, Sandel's analogy simply does not work.[2]

Finally, Sandel's main point seems to pivot around how he feels about the embryo. According to Sandel, an oak tree is not the same kind of thing as an acorn because we feel much more strongly attached to the former than to the latter. This is revealed in Sandel's statements that "more than half of all fertilized eggs either fail to implant or are otherwise lost," and "the way we respond to the natural loss of embryos suggests that we do not regard this event as the moral or religious equivalent of the death of infants."[3] Sandel gets his facts wrong about the percentage of embryos that die, but his assertion that the lack of a funeral means that the embryo is not a human person is a troubling one. What is it about having a funeral after you die that makes you a person? Did people slaughtered by Saddam Hussein, Pol Pot, or Adolph Hitler and buried in mass graves have a funeral? Were they not human persons? Of course they were. Those embryos that die do so before we know anything about them or have a chance to become attached to them. Just because we do not hold a funeral for their demise is neither here nor there when it comes to their personhood.

I think Sandel's ethic gets even more sinister if we take it to its logical conclusions. My oldest daughter returned from a mission trip to Minneapolis, Minnesota. While there, she and her friends encountered a homeless lady who was afflicted with bipolar disorder. Her parents lived four doors down from her daily haunt, but never even came to talk to her or to invite her home. Previously, she was even in a coma for several weeks, but

2. George and Lee, "Acorns and Embryos," 90–100.

3. Sandel, "Embryo Ethics," 208.

her parents never came to see her or gave the slightest indication that they cared about her condition. This homeless lady's parents have completely abandoned their own daughter. They don't like her mental illness. In short, they don't "feel" like being her parents anymore. Under Sandel's ethical criteria, what these parents did was morally fine. Yet any parent worth their moral salt will tell you that their children are *always* their children, and this homeless lady's parents have failed in their most basic duty to their own daughter. Such neglect is an outrage, but if we take Sandel seriously, this lady's parents were morally upright in all they did. The consequences of Sandel's treatment are heinous, un-Christian, and unworkable for any society.

In the end Sandel has offered a troubling recipe for justifying the destruction of embryos—because he does not feel a strong attachment toward them. Such thinking is not only unconvincing, but morally dangerous.

That's what I think. Keep working on those three-point shots. You're getting better.

Warmly,

Dr. Michael Buratovich

Letter #15

Human Is as Human Does

Dear Dr. B,

This whole stem cell thing has me stumped. Everywhere I look I see us evaluating entities based on what they do. In class you have emphasized that our value as humans is based on what we are and not on what we do. The problem is that I cannot for the life of me see how we can ever know what something is without seeing what it does. Does that make sense? I might think that someone or something is human, but until I see it do something human how can I possibly know that it is human?

This being the case, it seems to me that an embryo cannot be a human being. It can't think, reason, or argue, nor can it desire, love, hate and so on. Therefore it's not human.

Hit me with your best shot.

Nick H.

Letter #15

Dear Nick,

Thanks for asking a question that has been asked several times by many of my students, and others as well. Ideas have consequences and bad ideas have bad consequences. This is one idea that has some very bad consequences.

The problem with your statements is that they presuppose that "being a person" means "functioning as a person." This assumption doesn't work because there are times when you are not functioning, but you are still you. Are you still yourself when you are in a deep sleep? Does your identity change if you were to fall into a coma? How about when you were an infant and could not do what you can presently do: were you still you at that stage, or did you become you when you acquired all your present capabilities? How about when you get older and start to lose the ability to do some of the things that you can do now—will you stop being you? Nick, our capabilities change with age but our identity does not. Our identity stays with us no matter what we can or cannot do.

Foundational to the question that you have asked is a philosophy called functionalism. Functionalism muddles the relationship between function and essence. It confuses effects with the thing that produces those effects. What we can do is simply an effect of being a human person, but we are able to do these things because we ARE human. In other words, we have a human nature, essence, or being, and this allows us to do the sorts of things that humans do. Functionalism gets the cart before the horse. You do not do something in order to be someone; instead, you must be someone first before you can do something.

Consider the options for defining personhood if you adhere to functionalism. Is a person one who is consciously performing various human acts? Then this definition excludes people who are asleep, and thus killing someone who is asleep is not immoral, which is clearly absurd. What if we say that those who possess a present capacity to perform human functions are persons? This would include sleeping people, but it would exclude people in comas. Can we kill people in comas without guilt? Certainly not. What if we say that persons are those who have a history of performing particular human acts? This runs into problems because it generates absurdities. What if a twenty-five-year-old who was born in a coma twenty-five years ago wakes from her coma? She would not be a person under this definition, which is absurd. This also seems to suggest that you cannot do human acts without some kind of previous human act, which is to say that there can be no first human act. Thus this definition does not work either. What if we then say that a person is someone with a future ability to perform specifically human acts? Now those who are dying are not persons. None of these options for defining a person from a functionalist perspective work.[1]

There is another nasty application of functionalism: there are very few differences between a baby in the later fetal stages and one who has been newly born. If functional criteria are the standards you are using to determine if someone is or is not a person, then why don't you favor infanticide? This is the logical extension of functionalism. If you embrace it, then you need to consider this chilling consequence.

The bottom line is that a person is someone who has an inherent, natural capacity for performing specifically human acts. Embryos are persons because they grow into this ability to perform these acts. They can grow into these abilities because they are already persons. Only persons are the kind of thing that grows into such abilities.

That's my take on this one.

Michael Buratovich

1. Kreeft, *Three Approaches to Abortion*, 96.

Letter #16

Doesn't the Embryo Gradually Acquire More and More Rights as It Develops?

Dear Dr. Buratovich,

My sister is in your senior seminar class. I had a question about the stem cell debate, and after talking to my sister, she suggested that I ask you. I know you are busy, but could you please spare me a minute to answer my question?

My sister came home and talked about how a fertilized egg is a human person. However, my understanding of human development is that it is a continuous process that does not allow one to draw thick lines of distinction between when someone becomes a person. Since development is a process and not an instant event, then how can you pin down fertilization as THE event that makes a human person? Doesn't a person gradually become a person as they gradually develop?

Can you address my question, please?

Thank you for your time.

David C.

Doesn't the Embryo Gradually Acquire More and More Rights as It Develops?

Dear David,

Thank you for your e-mail. I am grateful that your sister turned you on to the stem cell issue. You have asked a very good question, and I hope I can give you a satisfactory answer.

You have hit upon a theory of human personhood known as gradualism. Gradualism understands the moral worth of an embryo or fetus as "proportional to the gradual actualization of certain potentialities relevant to the determination of moral worth or personhood."[1] This view, however, depends upon a functionalistic definition of human personhood, since it uses the acquisition of abilities and physical features to gauge the personhood of the embryo. Some people prefer the gradualistic view of human personhood because, they might argue, fertilization and early cleavage are not discrete events, but part of a continuous and sometimes overlapping symphony of occurrences that defy specification of one particular event as the point at which human personhood begins. For example, bioethicist Ronald Green writes, "Biology does not admit of definitive events. Instead it almost always involves complex processes with many occurrences and transitions happening over periods of time."[2]

Green, however, goes even further and argues that because the continuity of human development does not present a definitive stage in which embryos or fetuses become human persons, it is up to each individual to decide when an embryo becomes a person and then treat it accordingly: "It is not just a matter of discovering important events in the entity that must dictate our judgment. Rather, identifying these events requires us to identify and apply the values that underlie our thinking. Drawing on these values, we must decide which events are most important to us among the range of alternatives."[3]

Thus there is no objective standard for addressing the personhood of individuals, only the value judgments and perspective of the one deciding. Robert P. George and Christopher Tollefsen point out the fallacy of Green's position:

> [I]f the identification of biological events is a matter of decision, and not fact, then there is nothing in principle to prevent two people, or two groups, from holding radically different but equally valid positions about when something had happened. . . . But any view that holds, in regard to some type of claim or other, that there

1. Hui, *At the Beginning of Life*, 322.
2. Green, *Human Embryo Research Debates*, 32.
3. Ibid.

> are no facts, no "right answers," but that answers, knowledge, or truth are merely a matter of decisions, radically removes the possibility of error by making virtually every answer a right one.[4]

Likewise Green and others like him are almost certainly mistaken when they assert that the continuity of human development prevents determining when human personhood begins. One could just as easily see a road trip from Detroit, Michigan, to Los Angeles, California, as a continuous venture, but it is easy to identify events that mark when one begins and ends the trip. In the same way, even though fertilization and the onset of the first cleavage are fairly continuous, when the sperm and egg fail to exist as distinct entities, fertilization is complete and cleavage ensues. No new genetic information is required for human development to continue, and even though the embryo must implant into the mother's endometrium to continue further development, we are all dependent on others for our lives. To extend a lack of physical dependence upon the embryo as a requirement for personhood is unreasonable.[5]

There is, however, a more insidious side to gradualism. In order to assert that the human embryo acquires human personhood you must also assert that the embryo becomes more of a person as it acquires particular abilities. In other words, you need "a criterion of moral standing."[6] Therefore gradualism grades the embryo according to what it is able to do. That is functionalism, pure and simple. I have ranted and raved against functionalism in class several times, and the objections that apply to functionalism apply equally well to gradualism, since gradualism is largely applied functionalism.

Westmont University emeritus professor of philosophy Robert Wennberg argues for a gradualistic view of acquisition of personhood by asserting a right to life at conception leads to "certain oddities." For example, "the use of the intra-uterine (IUD) and the so-called morning-after pill, both of which serve to induce abortion, is as serious a moral offense as causing the death . . . of a seven-month-old fetus, a newborn infant, a child, or an adult."[7] He continues by arguing that "most people" regard the use of an IUD "more like contraception in its moral seriousness than like infanticide, manslaughter, or possibly even murder."[8]

4. George and Tollefsen, *Embryo*, 126.
5. Beckwith, *Defending Life*, 160.
6. Lee, *Abortion and Unborn Human Life*, 48.
7. Wennberg, *Life in the Balance*, 67.
8. Ibid.

Doesn't the Embryo Gradually Acquire More and More Rights as It Develops?

This is an argument built almost entirely on intuition, which is dangerous because people's intuitions not only differ drastically in some cases, but they are often wrong. I once held a large weight suspended from a string that swung as a pendulum and had my students stand with their chins to the weight. Once the weight was let loose to freely swing, we watched to see if the student moved back. The physics of pendulums clearly showed that the student had nothing to fear, but their intuition consistently caused them to jump backwards. Their intuition, you see, was simply wrong in this case. Moral intuitions are no different. Would we use the moral intuition of most people regarding slavery during the eighteenth century to assess the morality of slavery? I hope not. Secondly, people can become inured or numb to the wrongness of particular acts, but this does not lessen the heinousness of those acts. If a human embryo is a very young human person, then killing it with an IUD or a morning-after pill is just as objectionable as killing the adult next to you, but the fact that we cannot see the death of the embryo prevents us from seeing it for what it is. In this case the moral intuition of "most people" is simply mistaken because of what they do not see. Also, the fact that Wennberg offers no statistical data to justify his notion makes his argument even less convincing.

There is also a somewhat troubling consequence of gradualism. If people gradually acquire human personhood at the beginning of life, can they gradually lose it as they gradually lose their faculties?[9] Wennberg insists that this need not be the case because once one has the right to life he or she cannot lose it, and we should treat senile people with the same "overflow" of compassion that we show to mentally deficient people.[10] If this is the case, then why don't we extend that overflow to the embryo? Wennberg's invocation of the lack of a loss of a right to life—even though his gradualism implies such a loss—is gratuitous.

In conclusion, gradualism relies on the failed philosophy of functionalism, is largely supported by unreliable intuitions, and has monstrous implications if relied upon. Therefore I reject gradualism as a means of determining the personhood of human embryos—and so should you.

Thanks again for your question. I hope to meet you one day.

Sincerely yours,

Michael Buratovich

9. Devine, *Ethics of Homicide*, 78.

10. Wennberg, *Life in the Balance*, 119–21.

Letter #17

"But They're Going to Die Anyway!"

Dear Dr. B bomb,

Thank you for your note. I am going to miss my brother, but he is going to serve his country in a dangerous place and I am proud of him.

While I have you online, could you answer something for me? In class last week, Josh brought up the millions of embryos sitting in cold storage, and he asked why these shouldn't be used for embryonic stem cell research. I understand that you think that these embryos are human beings and everything, but they are going to eventually die anyway in cold storage or their parents are going to flush them down the sink. Why not do something useful with them? To not do so seems to me to be a terrible waste. To serve humanity in research seems to be a much better way to die than death in the sewer or a life in the limbo of cold storage. Don't you think so?

Also, what about the reproductive technologies that produce all these extra embryos in the first place? You have not really said much about them.

Erin P.

Dear Erin,

This is probably one of the most popular arguments brought to bear in favor of embryonic stem cell research: "We have these embryos and they are under a death sentence. No one wants them and they are going to die ignominiously. To not use them for research is wasteful and unreasonable."

First of all, let me address one misconception. There are not "millions" of embryos sitting in cold storage in fertility clinics around the United States. In 2002 the Society for Assisted Reproductive Technology and the RAND Corporation conducted a study of 430 fertility clinics in the United States. They determined that there were 396,526 embryos in cryostorage as of April 11, 2002. Today this number is almost certainly well over half a million and might even be close to one million (no one truly knows how many embryos are presently in cryostorage). Parents claimed 88.2 percent of these frozen embryos to grow their families. Of the remaining 47,000 embryos, 9,225 were intended for donation to other infertile couples, 8,840 were discarded, 18,000 were of an uncertain status (divorce, the clinic had lost contact with the parents, etc.), and 752 were destroyed because they were grossly abnormal or had stopped growing. Eleven thousand two hundred eighty-three embryos had been donated to embryo research, but an estimated 65 percent would survive the freeze/thaw process (leaving 7,334), and of these, only 25 percent would survive to the blastocyst stage (1,834). If we estimate a 7.5 percent to 27 percent success rate for deriving ESC lines, only 275 cell lines would result from the original 400,000 embryos.[1] Therefore, there are not "millions" of embryos available for embryonic stem cell research in the United States, and even if we gave all the available embryos to researchers, they would not make thousands of embryonic stem cell lines, but only a maximum of a few hundred.

In Canada, a similar study showed that 15,615 embryos were currently in storage in fertility clinics. Of these, less than 2 percent (299) had been donated for research.[2] Another 2002 study suggests that there are as few as 500 embryos available in Canada for research.[3] Once again, we see that there are nowhere near a million embryos available for research.

Secondly, the claim that "they are going to die anyway" is an exceedingly poor justification for embryo-destructive research. If this justification applies to the killing of human persons, then why do we not perform

1. Hoffman et al., "Cryopreserved Embryos," 1063–69.
2. Baylis et al., "Cryopreserved Human Embryos in Canada," 1026–31.
3. Baylis, "Human Embryonic Stem Cell Lines," 159–63.

life-threatening research on death row inmates? Such a suggestion would be justifiably labeled as cruel and barbaric. However, with very young human persons it is somehow justified? This makes no sense. Essentially the argument assumes the nonpersonhood of the embryo and then makes its case based on that presumption.

This is well illustrated by a story told by Scott Klusendorf. Let's pretend that you oversee a Cambodian orphanage with 200 abandoned toddlers, and your circumstances are dire. Your levels of potable water are critically low and all food stores have been exhausted. It is only a matter of time before starvation and disease set in. A foreign scientist comes to you with an offer to take the toddlers off your hands for use in gruesome medical experiments that are designed to find a cure for cancer. He reminds you that many of these children will die soon and there is not a thing you can do to prevent it. Why then should you let all those orphans go to waste? You refuse, and you would never, even for a moment, consider giving your orphans over to this scientist on grounds that "these kids are going to die anyway so let's put them to good use." Given your impoverished circumstances, you are powerless to save them, but you would never be complicit in actively killing vulnerable human beings.[4] However, the popular justification for embryonic stem cell research concludes that it is perfectly fine to give these children over to the doctor because "they're going to die anyway." It is a simple fact that we are all going to die anyway, and those of us who are going to die later have no moral right to exploit those who will die sooner.

The question comes back to the nature of the embryo, and the fact that the embryos are going to die sooner rather than later is completely immaterial. If the embryo is a human person, then there are really only two acceptable moral choices, and those are to implant them yourself or give them to adoptive parents.

With respect to your second question, there is much to say about the production of these extra embryos. The procedure known as "in vitro fertilization" or IVF is one of a group of procedures collectively known as artificial reproductive technologies (ARTs). Of the ARTs, IVF is by far the most commonly used and generates the vast majority of extra embryos.

Thirteen to fourteen percent of couples at the childbearing age suffer from infertility. Formally, infertility is defined as the inability to conceive after one year of regular, unprotected sexual intercourse. There are a host of

4. Crossway, "Scott Klusendorf on Embryonic Stem Cell Research."

reasons couples might be infertile, and there are several treatments for it. However, when all other treatments fail, couples turn to IVF.

IVF procedures begin with "controlled ovarian hyperstimulation" protocols that amp up the egg-producing capacities of the mother's ovaries. These protocols require the future mother to inject herself with particular hormones on a strict schedule. If these hormones successfully stimulate her ovaries, she follows up these injections with the self-administration of a hormone called human chorionic gonadotropin hormone (hCG) to mature the eggs in the ovaries so that they are almost ready to undergo ovulation. Approximately thirty-six hours later, an aspiration needle is passed through the wall of the mother's birth canal, into her abdomen to collect the mature eggs from the ovaries.[5] The retrieved eggs are mixed with sperm collected from the father in a culture dish, and the resultant embryos are grown in culture for two to five days and then transferred into the uterus of the mother for implantation, or frozen (cryopreserved) for later pregnancy attempts (fig. 17.1).

5. Goldberg et al., "In Vitro Fertilization Update," 329–38.

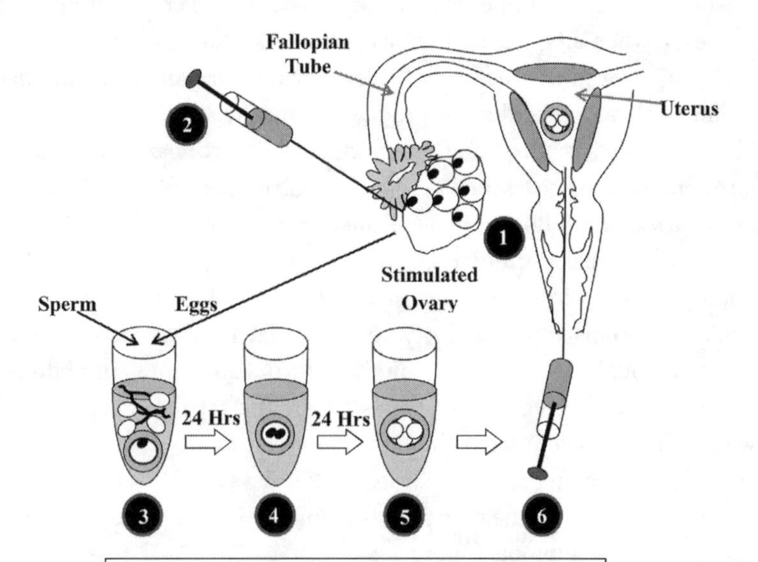

Figure 17.1 In Vitro Fertilization Procedure

1. Stimulation of the ovary through administration of drugs that cause maturation and ovulation of multiple eggs at once.
2. Collection of eggs by means of transvaginal aspiration.
3. Exposure of the collected eggs to sperm in a test tube.
4. After 24 hours, zygotes are formed and cleavage stages have begun.
5. 48 hours later the embryo is at the four-cell stage.
6. Embryo transfer into the mother's uterus.

The success rate of IVF varies from clinic to clinic and from couple to couple, but the age of the mother is the most important variable.[6] According the U.S. Centers for Disease Control and Prevention (CDC), women under the age of thirty-five have an average success rate of 39 percent, women aged thirty-five to thirty-seven have an average success rate of 30 percent, women between the ages of thirty-seven and forty have an average success rate of 21 percent, and women aged forty-one or forty-two have an average success rate of 11 percent.[7] The CDC also tabulates the number of IVF births in the United States, and in 2008 there were 148,055 ART *cycles* or

6. Klonoff-Cohen and Natarajan, "Advancing Paternal Age," 507–14.
7. Centers for Disease Control, *2006 Assisted Reproductive Technology Success Rates*.

treatments that resulted in 46,326 live births that produced 61,426 infants. Clearly this is a heavily used technology.[8]

Presently in the United States biotechnologies that affect human reproduction receive minimal direct governmental regulation. Furthermore, a 2004 report issued by the President's Council on Bioethics found that no uniform, comprehensive, or enforceable system of data collection, monitoring, or oversight of ARTs presently exists. While some degree of self-regulation exists among fertility specialists, compliance with these published guidelines and recommendations is completely voluntary.[9] No restrictions exist on the number of embryos conceived, and this leads to greater numbers of extra embryos. The present regulatory situation in the United States is singularly unwise and is not a good example of legislation that accounts for the value and human personhood of the embryo.[10]

In addition to being poorly regulated, IVF also poses risks to mothers, babies, and embryos, and imposes particular stresses upon parents and society in general.

The first IVF risk consists of the fertility drugs the mother takes to stimulate her ovaries to release large quantities of eggs. In about 10 percent of all cases, these drugs can cause ovarian hyperstimulation syndrome (OHSS), in which large amounts of fluid become shifted from the rest of the body to the ovaries.[11] Severe OHSS can cause blood clots, kidney failure, fluid accumulation in the abdomen or chest, and severe salt imbalances; in some cases it can even cause death.[12]

The second risk is multiple births. ARTs are associated with higher rates of multiple births (twins, triplets, and so on). Louise Joy Brown, the first baby conceived by IVF, was born on July 25, 1978. After this time, IVF became more and more heavily used to treat infertility. In the United States, the rates of twins increased by 75 percent from 1980 to 2000,[13] and similar trends occurred in Europe.[14] Infertility treatments account for 30 percent to 50 percent of all twin births and more than 75 percent of all high-order multiple births (triplets, quadruplets, and so on).[15] Multiple births increase

8. *2008 Assisted Reproductive Technology Success Rates.*
9. President's Council on Bioethics, *Reproduction and Responsibility.*
10. Hook, "In Vitro Fertilization and Stem Cell Harvesting," 282–89.
11. Whelan and Vlahos, "Ovarian Hyperstimulation," 883–96.
12. Fineschi et al., "Fatality Due to Ovarian Hyperstimulation," 293–99.
13. Hogue, "Beauty of One," 1017–19; Jones, "Multiple Births," 17–21.
14. Bréart et al., "Childbearing Population in Europe," S45–52.
15. Fauser et al., "Multiple Birth from Ovarian Stimulation," 1807–16.

the risks for a variety of complications for mothers and their babies. Mothers who give birth to multiple babies suffer from higher rates of high blood pressure during pregnancy, higher rates of cesarean section, more bleeding after delivery, prolonged bed rest due to premature labor, and, although rare, maternal deaths. Babies born during multiple births have increased rates of low birth weight, premature birth, respiratory problems, cerebral palsy, deafness, blindness, and death.[16] Anyone who opts for IVF should be aware of these risks.

Third, IVF presents risks to embryos, since it depends upon embryo freezing, which is potentially deleterious to them. Despite improvements in the embryo freezing procedure (cryopreservation), rates of embryo survival after freezing range from 70 percent to 80 percent.[17] Furthermore, cryopreservation and thawing can damage embryos. Thus even if an embryo survives the freeze/thaw process, the damage it experiences can lead to its premature demise.[18] Therefore, IVF depends upon a procedure that potentially destroys or fatally damages embryos.

Fourth, the extra embryos created by IVF exert emotional stresses upon the parents. Surveys of IVF patients who have surplus embryos stored in deep freeze repeatedly demonstrate that these couples feel a tremendous amount of anxiety over what to do with their embryos. This stress tends to lead couples to put off making a decision about the destination of their frozen embryos. An American study showed that 72 percent of interviewed couples who had stored their frozen embryos for an average of 4.2 years had not reached a decision about what to do with them. Essentially, when couples treated by IVF have completed their families, they largely avoid the question of what to do with their embryos.[19]

Other studies show that this phenomenon is not restricted to American IVF patients. An Australian study revealed that 70 percent of women who had surplus frozen embryos were not able to come to a decision about their fate five years after having given birth to their first IVF-conceived baby. This same study revealed that couples deemed their decision regarding the fate of their embryos as the most difficult decision they have ever had to

16. Ombelet et al., "Multiple Gestation and Infertility Treatment," 314.

17. Michelmann and Nayudu, "Cryopreservation of Human Embryos," 135–41.

18. Guerif et al., "Selection of Best Embryos for Transfer," 1321–26; Barratt et al., "Challenges Providing Embryos for Stem-Cell Initiatives," 115–18; Bankowski et al., "Social Implications," 823–32.

19. Nachtigall et al., "Parent's Conceptualization of Frozen Embryos," 431–34.

make.[20] Other studies showed that 35 percent of Australian patients said that they had some concern about what to do with their frozen embryos, and 12 percent had a great deal of concern.[21] Those who participated in another Australian study described their feelings about their impending decision about the fate of their frozen embryos as "anguished" and "agonizing."[22] An Italian study also showed that 25.1 percent of Italian patients allowed their embryos to be destroyed without making a definitive decision.[23]

In order to prevent couples from indefinitely delaying their embryo-based decisions, many fertility clinics have instituted an advanced directive that is signed by the couple before they begin IVF treatments. However, these directives do not work because couples tend to change their minds when it comes to their decisions about their embryos. Research by Susan Klock and colleagues showed that only 29 percent of IVF patients adhered to their original decision as recorded in the advanced directive, and of those who chose to donate their embryos to research, 87 percent changed their minds.[24] A Canadian study found that at least 59 percent of all IVF patients changed their minds.[25] Therefore, published surveys that assert that particular percentages of IVF patients will donate their frozen embryos to research should be treated with some skepticism.[26]

Christians struggling with infertility can consider IVF as an option, but with several caveats. Here are some guidelines for those who opt for IVF, but wish to respect the human personhood of the embryo:

- Consider IVF when all other options have failed.
- Strongly consider adoption before you consider IVF. Talk to couples who have done both if you can. With the Internet, it should not be difficult to locate such couples.
- To date, the preservation of eggs (oocytes) is considered an experimental procedure, but because the protocols for oocyte preservation are improving every year, it is entirely conceivable that ART clinics

20. McMahon et al., "Embryo Dilemmas after Five Years," 131–35.

21. McMahon et al., "Embryo Donation for Medical Research," 871–77.

22. de Lacey, "Why Patients Discard Unused Embryos," 1661–69.

23. Cattoli et al., "Fate of Stored Embryos," S168.

24. Klock et al., "Disposition of Frozen Embryos," 69–70.

25. Newton et al., "Changes in Patient Preferences," 3124–28.

26. Lyerly and Faden, "Embryonic Stem Cells," 467; Cortes et al., "Spanish Stem Cell Bank," 17–20; Lyerly et al., "Views about Frozen Embryo Disposition," 499–509.

will have this procedure available as an option in the near future.[27] If you have the option, preserve oocytes and sperm but not embryos, except under exceptional circumstances.

- Implant every embryo the lab conceives.
- Insist on single embryo transfer.
- Decide beforehand that you will raise every single baby that results from each successful pregnancy, and that you will not submit to "fetal reduction."

Beyond that, the Christian church also needs a better theology of infertility. I have a few suggestions. First of all, we should argue that the Bible unabashedly endorses parenthood. God also sent his Son to redeem us by becoming a human being, but he did not beam down as an adult. Instead, he was born to a woman and was raised by two married parents (Matt 18:1–6; Mark 10:13–16; Gal 4:4). This is a direct endorsement of having childbirth and of parenting. Also, the Bible refers to God as our Father and to us as His children (Rom 8:16–17), which is yet another endorsement of having children. In fact, the Bible directly commands humanity to raise children (Gen 1:28), which provides a scriptural mandate for parenthood. Likewise, the Bible clearly regards children as a blessing and not a burden (Ps 127:3–5).

However, the Bible also teaches that we should not be presumptuous about what God gives us (Jas 4:15). Just because God blesses many couples who conceive and have children does not mean that God will or is under any obligation to bless all couples in this manner. Also, when it comes to bearing children, the Bible clearly argues that God opens and shuts the womb. The bearing of children is subject to the mysterious providence of God (1 Sam 1:5–6). Finally, God has a purpose for couples who are fertile and a purpose for couples who are infertile. He has a purpose for all Christians (Rom 8:28). Therefore, infertile couples should never be treated as second-rate Christians.[28]

This is probably a much longer answer to your question than you wanted, but you asked and this is what I think.

Keep the Faith,

Michael Buratovich

27. Van der Elst, "Oocyte Freezing," 463–70; Noyes et al., "Oocyte Cryopreservation," 69–74.

28. Kilner and Mitchell, *Does God Need Our Help?* 149–53.

Letter #18

Can We Use "Somatic Cell Nuclear Transfer" to Make Embryonic Stem Cells?

Dr. B.,

I wrote to my congressperson about the cloning issue. I have forwarded her response in this e-mail. She says that the bill that she voted for does not legalize cloning but does legalize "somatic cell nuclear transfer," which will allow scientists to make cells for lots of interesting cures. She also states that the bill will prevent "reproductive cloning." I guess she answered my question, but I am still a little puzzled. Isn't this exactly what we talked about in class? Now I'm confused. Can you translate this for me?

Thanks,

Michelle E.

Dear Michelle,

You aren't the only one who's confused after reading her response. Somatic cell nuclear transfer is cloning. I think I know what she is getting at, but I will try to show you why I find her answer rather unsatisfying.

Scientists have suggested that cloning is a way to make embryonic stem cells that are genetically identical to the cells of those patients who need them. Let's say that a patient has congestive heart disease, kidney failure, cirrhosis of the liver, or other diseases associated with failing organs. The best way to treat such degenerative diseases is to regenerate the affected organ with stem cells. However, the available embryonic stem cell lines would not match the tissue type of the patient, and if they were transplanted, the patient's immune system would promptly attack and reject them.

This is the part I didn't get to in class. The surface of just about every cell in our bodies is marked with a group of proteins that act as personal "bar codes." The immune system attacks or disregards transplanted cells by reading those bar codes and determining whether they match those found on cells in the rest of the body. These specific surface proteins that the immune system examines are called "histocompatibility antigens." Histocompatibility antigens act as the bar codes that our immune system recognizes and uses to determine what cells belong to our bodies and what cells do not. The immune system constantly surveys the proteins on the surfaces of our cells, and if they find cells that have the proper types of histocompatibility proteins on their surfaces, they typically leave them alone. If, however, the histocompatibility proteins are unlike those found in the rest of the body, then the immune system attacks those cells and destroys them. This explains transplantation rejection. Stem cells, if they are to be used for regenerative medicine, must possess the same histocompatibility proteins on their surfaces as those found on the surfaces of the patient's cells. If they don't, the immune system will attack them and destroy them.

Initial work on embryonic stem cells suggested that they might represent a special group of cells that were ignored by the immune system.[1] However, further work has clearly demonstrated that this is not the case.[2] Embryonic stem cells made from cloned embryos, however, would possess the same genes as the patient who donated the cells for cloning. Therefore, they would demonstrate the same bar codes and have a much lower poten-

1. Li et al., "Immuno-Privileged Properties," 448–56.

2. Drukker et al., "MHC Proteins in Human Embryonic Stem Cells," 9864–69; Drukker et al., "Embryonic Stem Cells Less Susceptible to Immune Rejection," 221–29.

tial for rejection by the immune system. Exceptions to rejection by the immune system are transplantations into the central nervous system because the blood-brain barrier separates the immune system from the brain and spinal cord.

Cloning involves the transplantation of the nucleus, that cellular compartment that houses the chromosomes, from a body cell other than an egg or sperm cell (a somatic cell) into an unfertilized egg whose nucleus has been removed. This egg, whose nucleus is now from a somatic cell, is activated by either the application of particular chemicals or a small electrical current. This reprograms the nucleus, and the egg behaves like a zygote (single-cell embryo): cell division and early embryonic development ensue.[3]

This technique I just described is called "somatic cell nuclear transfer," or SCNT. This is what the congresswoman was referring to in her letter, but there's more. If SCNT is used to make a human or animal embryo that is transferred into the uterus of a surrogate mother and is eventually born, then that cloned embryo constitutes an example of "reproductive cloning." The objective of reproductive cloning is to make genetically identical animals. However, embryos made by means of SCNT can undergo another fate. Cloned embryos can develop for approximately five to six days, have their outer trophoblast cells removed, and have their inner cell masses cultured to establish embryonic stem cells. Such embryonic stem cells are called nuclear transfer embryonic stem cells or NT-ESCs (fig. 18.1). This strategy is called "therapeutic cloning" because its goal is to form embryonic stem cells that are genetically identical to the patient who donated the nucleus. Because NT-ESCs are genetically identical to the patient, they should have a lower risk of being rejected by the immune system if they are used in regenerative medicine.

I think the congressperson meant that her bill outlawed reproductive cloning, but allowed therapeutic cloning. Herein lies why she said that SCNT was still allowed, but gestating a cloned embryo to birth was not.

Why do I have problems with this? First of all, treatments that use therapeutic cloning are fraught with problems. SCNT has been used to generate embryonic stem cell lines from laboratory mice with functionally deficient immune systems. These NT-ESCs were subsequently genetically engineered to repair the genetic defect that hamstrung the immune system, and then used to reconstitute the immune system of the parent mice.

3. Meissner and Jaenisch, "Mammalian Nuclear Transfer," 2460–69.

Unfortunately, the implanted stem cells still experienced rejection by the immune system.[4] Thus, even though the rationale for making NT-ESCs included avoiding rejection by the patient's immune system, they can still be rejected.

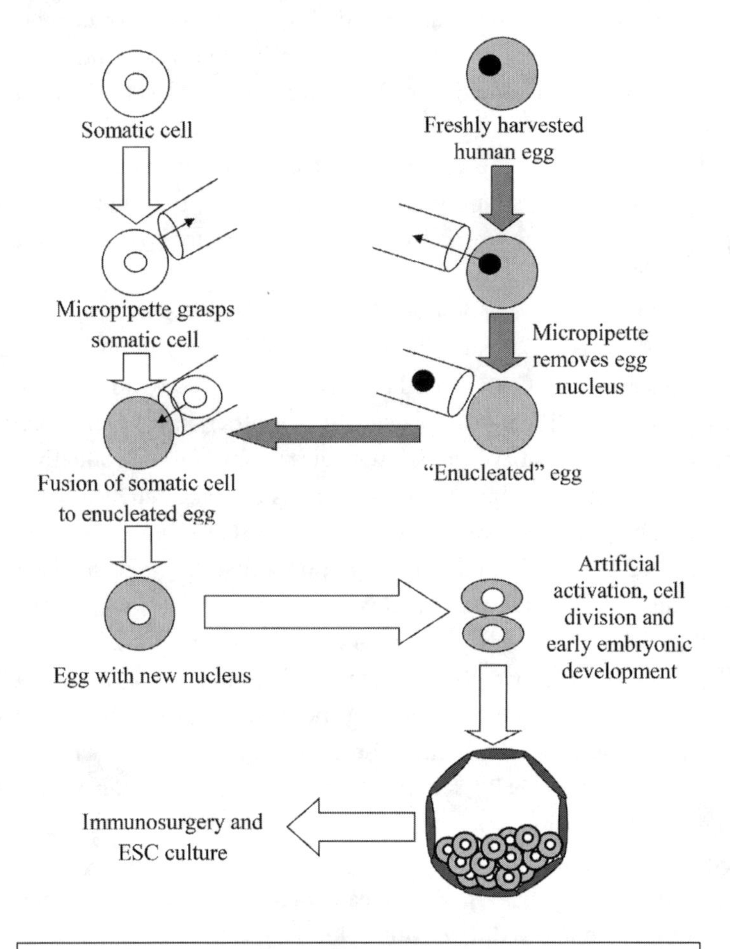

Figure 18.1 Therapeutic Cloning. This shows the protocol for somatic cell nuclear transfer. The somatic cell used here requires that it have a nucleus that has not been lost or degraded. If the blastocyst-stage embryo was implanted into the womb of a surrogate mother, it could ideally generate a newborn baby (reproductive cloning).

4. Rideout et al., "Correction of a Genetic Defect," 17–27.

Other therapeutic cloning-based treatments have experienced some successes. If NT-ESCs are used to treat the brain, which is isolated from the immune system by the blood-brain barrier, then such treatments are very successful.[5] Other successful uses of SCNT have been reported, but these experiments utilized tissue from cloned embryos that were gestated to the fetal stage, after which tissues from the aborted fetuses were harvested and used in successful transplantation experiments in adult animals. In July 2002, researchers at the Massachusetts-based company Advanced Cell Technology, after failing to isolate kidney cells from bovine (cow) embryonic stem cells, implanted cloned embryos into the uteruses of female cows. The implanted embryos grew to the fetal stage and were aborted; after the cloned fetuses were aborted, the fetal cow kidneys were harvested and successfully used in transplants in adult cows.[6] Similar experiments also produced successful transplantations of heart and liver tissue.[7] In a press release, Robert Lanza, medical director of Advanced Cell Technology, said, "We hope to use this technology in the future to treat patients with diverse diseases such as marrow failure disorders, various genetic diseases and malignancies, as well as debilitating autoimmune diseases, including MS, arthritis, diabetes, and lupus."[8] In December 2002, Israeli researchers reported that immature kidney cells, obtained from both a seven- to eight-week gestated human fetus and a three- to four-week gestated pig fetus, could give rise to a functioning kidney. The authors of this report concluded that this presented "a window of human and pig embryogenesis that may be optimal for transplantation in humans."[9]

Such usages show that SCNT can generate transplantable material, but only if the embryos produced in this manner are allowed to develop to the fetal stage and are then aborted. Growing fetuses for the sole purpose of killing them and using their tissues is called "fetus farming," which, for now, is illegal in the United States since the passage of the Fetus Farming Prohibition Act of 2006 (S.3504).[10]

5. Tabar et al., "Therapeutic Cloning in Parkinsonian Mice," 379–81.

6. Lanza et al., "Histocompatible Tissues Using Nuclear Transplantation," 689–96.

7. Lanza et al., "Stem Cells Derived by Nuclear Transplantation," 820–27. Lanza et al., "Clone-Derived Stem Cells," 95–106.

8. Griffin and Gambill, "Somatic Cell Nuclear Transfer."

9. Dekel et al., "Kidney Precursors for Transplantation," 53–60.

10. Fetus Farming Prohibition Act of 2006, Pub. L. No. 109–242, U.S. Statutes at Large 120 (2006) 570.

Think of it, Michelle—growing babies just to harvest their organs and grow stem cells from them, all because we cannot get this technology to work any other way. It sickens me to think that people might have children just to kill them so that they can save their own skins. What if we wanted the babies for beauty treatments instead of cures—would that be all right? This devaluates human life in the extreme. It also shows that there is no real difference between therapeutic and reproductive cloning—only the length of time the individual lives, and that is completely negotiable depending on the baby's utility to us.

Secondly, human cloning experiments require large quantities of human eggs. Robert Lanza made it clear when he said, "Without eggs, there is no research."[11] In order to acquire eggs for cloning, researchers look to young, fertile women to donate their eggs.

Fertility clinics have utilized egg donors for several years in order to help couples conceive and have babies. According to the Centers for Disease Control and Prevention, 11.4 percent of all procedures that involved assisted reproductive technology (like in vitro fertilization and related techniques) utilized donated eggs. In the United States, fertility clinics pay donors who provide eggs or sperm for reproductive purposes. Such payments are banned in Canada and the United Kingdom. In the case of egg donation, young, healthy women can make $15,000 or more per donation cycle.

In addition to fertility clinics, egg donations from young women are also needed by embryonic stem cell researchers for therapeutic cloning. However, the bioethical principles that regulate research at many large universities tend to categorize egg donation for research as a form of organ donation, which should be altruistic and free of financial incentives. Therefore, most research institutions prohibit payment for eggs donated for research purposes beyond expenses incurred during the donation process. For example, in its "Guidelines for Human Embryonic Stem Cell Research" published in April 2005, the National Academy of Sciences recommended that payments not be provided for egg donations. Likewise, the state of Massachusetts and the California Institute for Regenerative Medicine limit payments for research egg donations to expense reimbursements.[12] Supporters of this policy also believe that financial compensation might

11. Gruen, "Oocytes for Sale?" 285–308.

12. Steinbrook, "Egg Donation," 324–26.

disproportionately entice disadvantaged women to provide their eggs for research, making young girls a commodity to be bought and sold.[13]

Some bioethicists, however, have strongly argued that egg donors should be compensated for enduring such an arduous and complex procedure.[14] Egg donors spend "56 hours in the medical setting, undergoing interviews, counseling, and medical procedures related to the process."[15] After a full physical and psychological examination, the woman, if she is mentally and physically fit to be an egg donor, is provided with daily injections of synthetic hormones for seven to ten days.

The three-drug regimen for female egg donors includes drugs that mimic gonadotropin-releasing hormone (GRH), others that imitate follicle-stimulating hormone (FSH), and finally human chorionic gonadotropin hormone (hCGH). GRH and substances that imitate it put the woman's body into artificially induced menopause.[16] It puts the ovaries at "ground zero" so that all the eggs can begin maturation at once. The next drug regimen, FSH-imitating drugs, hyperstimulates the ovaries and induces the maturation of several egg-containing follicles at once. Typically, during a woman's menstrual cycle only one follicle releases an egg at a time. Treatment with artificial GRH and then FSH causes the maturation of several follicles at once so that several eggs are mature simultaneously.[17] Finally, a single injection of hCG triggers the release or ovulation of these eggs thirty-four to thirty-six hours after injection.[18] Mature eggs are retrieved by a brief surgical procedure in which, while the woman is under sedation, an aspiration needle is inserted through the vaginal wall under the guidance of ultrasound. Once in the abdominal cavity, the needle suctions the eggs from the mature follicles in the ovaries.

In addition to the tedium and complexity of this procedure, egg retrieval comes with some risks. Egg donors commonly experience

13. Dickenson, "Commodification of Human Tissue," 55–63; Check, "Research Volunteers or Organ Donors?" 606.

14. Thompson, "We Should Pay for Egg Donation," 203–9.

15. Ethics Committee of the ASRM, "Financial Incentives," S240–44.

16. Commercially available forms of GRH-imitating drugs include Lupron, Nafarelin, Triptorelin, Synarel, Prostap, Buserelin/Suprefact, and Goserelin/Zoladex. Some of these medicines come as a nasal spray, and Lupron can be administered as a single, large shot (Depot Lupron).

17. Commercially available forms of FSH (otherwise known as human menopausal gonadotropin, or hMG) include Gonal-F, Pergonol, Humegon, Menagon, Clomid tablets, and Urofollitropin.

18. Commercial forms of hCG include Pregnyl, APL, and Oxidrel.

abdominal swelling, tension, and pressure where their ovaries are located, mood swings, and bruising at the injection sites. GRH-like drugs also produce the symptoms of menopause, which include hot flashes. In fact, one particular GRH-like drug, leuprolide (Lupron), is not FDA approved for ovary hyperstimulation and is used off label for such procedures. In some women leuprolide causes horrific side effects like suicidal depression, racing heart rate, grand mal seizures, joint pain, blinding migraine headaches, and extensive hair loss. Even long after this drug is discontinued, some women experience numbness, joint pain, and memory loss.[19] Failure to abstain from sexual intercourse during the course of egg donation can result in unintentional pregnancies. However, these side effects pale in comparison to the really problematic side effect: ovarian hyperstimulation syndrome, or OHSS.

OHSS occurs in 1 to 10 percent of all egg donors, and its incidence depends on the drug regime used. In women with OHSS, blood fluids leak from the blood vessels and collect in the abdomen. This causes nausea, bloating, and even kidney damage. In severe cases, the lungs and space around the heart can fill with fluid, causing either heart stoppage or suffocation. Other consequences of severe OHSS include blood clots, dehydration, kidney failure, and twisting of the ovaries (adnexal torsion), which cuts off the blood supply and permanently damages them. Blood clots can cause heart attacks, loss of limbs, lung damage, strokes, and death.[20]

There are other risks as well. First, the long-term effects of the drug cocktails used on egg donors are uncertain. Some animal and epidemiological studies suggest that egg donors subjected to ovarian hyperstimulation have an increased risk of developing particular cancers,[21] even though there is no completely conclusive evidence that fertility drugs cause ovarian, uterine, or breast cancer.[22] Second, babies born to women who underwent ovarian hyperstimulation might also be at risk, since animal studies demonstrate that mice born to mothers whose ovaries were hyperstimulated showed increased incidence of low birthweight, growth retardation, and

19. James, "Painful Truth."

20. Practice Committee of the ASRM, "Ovarian Hyperstimulation Syndrome," 1309–14.

21. Glud et al., "Fertility Drugs and Ovarian Cancers," 237–57; Brinton et al., "Ovulation Induction and Cancer Risk," 261–74; Althuis et al., "Uterine Cancer after Clomiphene," 607–15; Pearson, "Health Effects of Egg Donation," 607–8.

22. Institute of Medicine, *Medical Risks of Human Oocyte Donation.*

cervical ribs (ribs sprouting from the portion of the backbone closest to the skull), a condition associated with stillbirth and cancer in humans.[23]

Because women who wish to donate eggs undergo a strenuous and burdensome procedure that may damage their health, few women are willing to donate eggs without financial compensation. The present shortage of human eggs for research underscores this dilemma[24] and has spurred many research institutions to rethink their egg donation policies. Furthermore, the analogy between egg donation and organ transplantation is not completely convincing, since organ donations have been conclusively demonstrated to save someone's life, whereas stem cell research presently can make no such claim.

Bioethicists and other professionals associated with biotechnology have begun to make a vigorous case for paying women who donate eggs for research. Debora Spar, professor of business administration at Harvard Business School, states that the United States "maintains the absurd inconsistency" that a woman can receive $20,000 for reproductive egg donations but nothing for research egg donations.[25] Bonnie Steinbock, a bioethicist at University of New York at Albany, notes, "Any time that we ask people to do things that impose significant burdens and some degree of risk, fairness may require that they be adequately compensated."[26] However, the hope that a small segment of young women will provide, at some risk, the raw materials for research free of charge is naive.

Payment for human eggs, unfortunately, has its dark side. Third-world women can be exploited for their eggs. Although their dire economic circumstances may compel them to donate, they are not informed of the medical risks, receive no follow-up medical care, and are paid a pittance for their egg donations. Romanian attorney George Magureanu represented two Romanian women who were victims of such "egg collecting factories." Both women, who were paid the equivalent of $250 for their donations, are exceedingly poor, poorly educated, and have lingering health problems because of their egg donations. The clinics refused to take any responsibility for their health problems.[27] Unfortunately, legal action against the clinics

23. Steigenga et al., "Increased Ribs after Ovarian Hyperstimulation," 63–68.

24. Maher, "Egg Shortage," 828–29.

25. Spar, "Egg Trade," 1289–91.

26. Steinbrook, "Egg Donation," 324–26.

27. Magureanu, "Human Egg Trading."

failed to yield justice for these two women, but it did make several nongovernmental organizations and the media aware of these events.[28]

There is no clear-cut way through this conundrum. Can research proceed with other sources of eggs? Some have suggested the option of using eggs from the ovaries of aborted fetuses.[29] Although aborting babies in the womb just for their tissue is barbaric, supporters of this policy might say the abortion occurred before the decision was made to use the eggs for research, and therefore the two decisions are not connected. Nevertheless, the fact remains that researchers would be benefiting from the induced death of the youngest members of the human race. Just because someone is dead does not mean that it is ethical to pick his or her pockets. Furthermore, eggs from fetuses are immature, and studies have shown that artificially matured eggs function poorly.[30] Another option is to use eggs made from embryonic stem cells. The problem here is the "chicken or the egg" problem: it takes eggs to make embryonic stem cells to make the eggs. Another option is to use animal eggs, which are quite abundant. However, the therapeutic utility of stem cells made from animal eggs is dubious, since they would contain animal DNA in various cellular compartments.[31] Such contamination could cause the immune system of the patient to reject them.[32] Finally, the least ethically contentious option is to use eggs from ovaries that were removed after surgery (ovariectomies).[33] This strategy would probably not provide sufficient quantities of eggs for research.

In conclusion, cloning degrades the humanity of women and commoditizes them.

Let me know if you have more questions I can address.

Michael Buratovich

28. Lundin, "Ethnographic Study on Fertility Tourism and Egg Trade," 327–44.

29. Hutchinson, "Aborted Foetus Could Provide Eggs?"

30. Eppig and O'Brien, "Mouse Oocytes from Primordial Follicles," 197–207; Dennis, "Synthetic Sex Cells," 364–66; Smith et al., "Effects of In Vitro Culture," 38–47.

31. Cells contain compartments called "organelles," or little organs. Organelles are committed to particular functions within the cell. One of these organelles, the mitochondrion, is the power generation organelle of the cell. Mitochondria contain their own DNA and gene expression/protein synthesis machinery.

32. Gruen and Grabel, "Roadblocks to Stem Cell Therapy," 2162–69.

33. Gruen, "Oocytes," 143–67.

Letter #19

Are Cloned Embryos Human Persons or Are They Only Manufactured Artifacts?

Dear Dr. B.,

When we talked about cloning this morning, I found your conclusions very unconvincing. It seems to me that a cloned embryo is a result of a laboratory manufacturing process, whereas embryos are the result of the natural process of fertilization. Cloned embryos are so radically messed up it seems to me that they are simply not worth getting choked up over them. Face it, cloned embryos can provide us with a fine source of embryonic stem cells without all the hassle over unused embryos. So I would say, "Clone away and let the stem cell isolations begin."

That's what I think.

Kyle T.

Dear Kyle,

Your argument resembles that of psychiatrist Paul R. McHugh. McHugh has argued that embryos made by means of somatic cell nuclear transfer (SCNT) are not human persons even though those made by fertilization are. According to McHugh, SCNT is a "biological manufacturing process" that is used to make not babies, but embryonic stem cell lines, and that "resembles tissue culture" more than fertilization. McHugh has even fashioned the name "clonotes" for SCNT-derived embryos to distinguish them from embryos made by fertilization with sperm.

McHugh's main argument that there is a definitive difference between embryos made by fertilization and those made by SCNT is as follows: "If one used the notion of 'potential' to protect cells developed through SCNT because with further manipulation they might become a living clone, then every somatic cell would deserve some protection because it has the potential to follow the same path."[1]

In other words, because nuclei from almost any somatic cell can be used to form a clonote, almost any somatic cell has the potential to become a clonote. It is absurd to regard all the somatic cells of our bodies as human persons. As Robert Lanza stated, "Research advances are making all cells embryonic, but if you consider these cells human life, then 100 souls are lost every time I sneeze."[2] Since it is untenable to regard somatic cells, which have the ability to form clonotes, as human persons, it is equally untenable to regard clonotes as human persons.

However, this analogy of somatic cells with embryos seems hopelessly flawed. In fact, the entire category of "potential embryos" is simply nonsensical. The term *embryo* refers to a very specific entity in the life of an organism. Something is either an embryo or not. Secondly, somatic cells are not similar to embryos. Instead, they are similar to sperm and eggs, the cells that are used to make embryos. Once the sperm and the egg fuse and complete conception, they no longer exist. In their place a new entity, the embryo, which did not exist beforehand, begins its existence. The embryo is a "distinct, complete, self-integrating organism."[3] Somatic cells are not organisms, but are, instead, part of an organism. An embryo may result from either SCNT or fertilization. Thus McHugh's first argument fails.

1. McHugh, "Zygote and 'Clonote,'" 209–11.
2. Fischer, "The First Clone."
3. George and Lee, "Acorns and Embryos," 90–100.

McHugh's second argument points to the fact that the vast majority of clonotes are grossly abnormal and die very early during development.[4] Since clonotes are not human persons, that makes the production of embryonic stem cells from them morally justifiable.

However, McHugh's second argument is flawed on four counts. First, while many cloned animals develop into animals with a variety of developmental abnormalities, not all of them do. To classify cloned animals as a distinct kind of creature because they possess abnormalities ignores those cloned animals that either do not possess such abnormalities or whose health is similar to animals that were not cloned. If cloned animals are assigned to a different category based on abnormalities, then that classification fails because some cloned animals are normal.

Second, molecular comparisons of cloned embryos with embryos made from in vitro fertilization show extensive similarities.[5] The abnormalities typically arise later, once the embryo implants into the uterus.[6] Thus the abnormalities that McHugh uses to disqualify cloned embryos as human persons have not yet occurred.

Third, if cloned embryos differ in kind from embryos made by fertilization, then what about cloned animals that survived to term? Are such animals a different kind of animal? Consider Dolly, the cloned sheep. Was she not a Finn Dorset sheep? If she was not a Finn Dorset sheep, then what was she? Some lower class of sheep? This simply makes no sense. If relegating cloned animals to a lower status is fallacious, then it is also just as fallacious to demote cloned human embryos to a lower, nonhuman status.

Finally, even if cloned embryos have abnormalities, so what? Do we really want to dismiss the humanity of individuals because they have some sort of handicap? Dismissing the humanity of cloned embryos because of their potential abnormalities is tantamount to dismissing abnormal

4. Jaenisch, "Testimony: President's Council on Bioethics," July 24, 2003; Humpherys et al., "Abnormal Gene Expression in Cloned Mice," 12889–94; Bortvin et al., "Incomplete Reactivation," 1673–80; Mann et al., "Disruption of Imprinted Gene Expression," 902–14; Rhind et al., "Human Cloning," 855–64; Kohda et al., "Variation in Gene Expression in Cloned Mice," 1302–11.

5. Somers et al., "Gene Expression Profiling," 1073–84; Yang et al., "Nuclear Reprogramming of Cloned Embryos," 295–302; Long et al., "Gene Profiling of Cattle Blastocysts," 243–56; Beyhan et al., "Transcriptional Reprogramming," 637–49.

6. Camargo et al., "Comparison of Gene Expression," 487–96; Arnold et al., "Somatic Cell Nuclear Transfer," 279–90; Wakayama, "How to Improve Cloning Efficiency?" 13–26; Bauersachs et al., "Endometrium Responds Differently to Cloned Embryos," 5681–86.

children and allowing medical research to be performed on them since their death is imminent.[7] Such a proposal is morally revolting.

Cloned embryos are human persons. Making and then destroying them for research is immoral. These are the reasons why I do not support cloning as a way to solve the stem cell ethics problem.

Keep thinking,

Michael Buratovich

7. George and Tollefsen, *Embryo*, 184–89.

Letter #20

Why Not Reproductive Cloning?

Dear Dr. B-Man,

In lab about two weeks ago we somehow got onto the subject of cloning and you told us about two types of cloning—therapeutic cloning and reproductive cloning. You said that the vast majority of scientists and science policymakers opposed "reproductive cloning" but were largely favorably disposed towards "therapeutic cloning." Kyle and Michelle asked a bunch of questions about therapeutic cloning and that discussion was really interesting, but I never really got to ask my question about reproductive cloning.

My question is simply this: What's so wrong with reproductive cloning? You said that many people oppose it, but some don't. What are the downsides and even upsides of cloning for babies?

Thanks for this one,

Brad S.

Dear Brad,

I'm glad you asked instead of letting this one go. I will give you the pros and cons of reproductive cloning as I see them and then tell you why I think the cons outweigh the pros.

The technology used for reproductive cloning is the same as that used for therapeutic cloning. Because we talked about this in class, I will not go over it again in this e-mail. Suffice it to say that somatic cell nuclear transfer (SCNT) experiments with several different animals have collectively revealed that cloning with either embryonic or somatic cells is highly inefficient. The success rate varies between different species and different laboratories and ranges from 0 percent to 7.2 percent.[1] Further work has not consistently raised the efficiency of this process.[2] To date, cloned adult sheep,[3] cattle,[4] goats,[5] mules,[6] horses,[7] pigs,[8] mouflons (a wild sheep),[9] mice, rats,[10] dogs,[11] cats,[12] and rabbits[13] have been made in the laboratory. While no cloned monkeys have yet to be born, cloned monkey embryos have also been made in the laboratory, and then used to generate embryonic stem cells.[14]

Many cloned mammals that are grown in surrogate mothers die before birth. The miscarried fetuses of cloned animals repeatedly show particular defects. Cloned animals that are born alive also tend to exhibit particular defects. Some of these defects are common to all cloned mammals, while others are specific to particular species. For example, cloned mice tend to show obesity, a feature not observed in other cloned mammals. The severity

1. Mombaerts, "Therapeutic Cloning in the Mouse," 11924–25; Rhind, "Human Cloning," 855–64.

2. Yang et al., "Nuclear Reprogramming of Cloned Embryos," 295–302.

3. Wilmut et al., "Viable Offspring," 810–13.

4. Galli et al., "Leukocytes Contain All Genetic Information," 161–70.

5. Baguisi et al., "Goats," 456–61.

6. Woods et al., "Mule Cloned," 1063.

7. Galli et al., "Cloned Horse," 635.

8. Polejaeva et al., "Cloned Pigs," 86–90.

9. Loi et al., "Genetic Rescue of an Endangered Mammal," 962–64.

10. Zhou et al., "Generation of Fertile Cloned Rats," 1179.

11. Lee et al., "Dogs Cloned from Adult Somatic Cells," 641.

12. Shin et al., "Cat Cloned by Nuclear Transplantation," 859.

13. Chesné et al., "Cloned Rabbits," 366–69.

14. Yang and Smith, "ES Cells Derived from Cloned Embryos in Monkey," 969–70.

of these defects varies between cloned individuals and between species. For example, defects in cloned mice, goats, and pigs are, on average, less severe than those in cloned sheep and cows.[15]

Some abnormalities that are somewhat common in cloned mammals include cardiovascular abnormalities, defects of the body wall, muscle, and skeleton, and large-offspring syndrome (a condition characterized by overgrowth of the body and organs; see table 20.1).[16] Large-offspring syndrome is not unique to cloned mammals, however, since it is also seen in animals whose embryos were subjected to different types of manipulations.[17] Cloned animals that die before birth frequently display placental deficiencies.[18] The table below illustrates the commonly observed pathologies of cloned animals.

Despite these observed defects, many cloned animals succeed in being born, and some make it to adulthood and seem rather healthy.[19] Surveys of the health of cloned animals show that they suffer from the same abnormalities as normally conceived animals, but at higher frequencies.[20] Thus, while cloned mammals are much more likely to suffer from particular developmental abnormalities, not all of them do.[21]

The production of cloned human embryos was reported later,[22] but the first reports of cloned human embryos by South Korean veterinarian Woo-Suk Hwang were completely fabricated.[23]

15. Rhind et al., "Human Cloning," 855–64.

16. Young et al., "Large-Offspring Syndrome," 155–63; Farin et al., "Development of Cloned Bovine Embryos," E53–62; Farina et al., "Errors in Development," 178–91.

17. Sinclair et al., "Aberrant Fetal Growth," 177–86.

18. Palmieri et al., "Pathology of Abnormal Placentae," 865–80; Hill et al., "Placental Abnormality," 1787–94; De Sousa et al., "Gestational Deficiencies in Cloned Sheep," 23–30; Tanaka et al., "Placentomegaly," 1813–21.

19. Lanza et al., "Cloned Cattle Can Be Healthy," 1893–94.

20. Cibelli et al., "Health Profile of Cloned Animals," 13–14.

21. Keefer et al., "Production of Cloned Goats," 199–203.

22. French et al., "Human Cloned Blastocysts," 485–93.

23. Buratovich, "Hwang Woo-Suk," 1094–97.

Table 20.1 Pathologies of Cloned Animals

Organ	Cattle[1]	Sheep[2]	Goats	Pigs	Mice
Placenta	Impaired development, swelling[3]	Reduced blood vessels[4]			Enlarged placenta[5]
Body Weight	Higher[6]			Lower[7]	
Heart	Right ventricle enlarged	Enlarged		Right ventricle enlarged[8]	
Lungs	Hypertension[9]	Hypertension, misaligned pulmonary vessels	Pneumonia[10]		Pneumonia[11]
Central Nervous System		Pathology[12]			
Kidneys	Abnormalities, including size[13]	Dilation, blockage in lower tubules[14]			
Blood System	Low red and white blood cell counts[15]				Immune system impaired[16]
Endocrine System	Diabetes				
Liver	Fibrosis, fatty liver[17]	Enlarged, few bile ducts, fibrosis			Dead liver cells[18]
Muscle/ Skeleton	Limb deformities[19]	Body wall defects			
Other					Obesity[20]

All this is to say that reproductive cloning is risky business, and making a human being in this way would probably produce sick individuals whose chances for survival were exceedingly poor. This is one very good reason not to clone people.

However, what if it were possible for scientists to completely work out the bugs in this procedure and make people who were not abnormal? Would that make it right? My answer would still be no, and here's why.

Think about why someone would want to have a child by means of reproductive cloning. One scenario we can imagine is that a child has died, and the parents opt for reproductive cloning to replace their dead child. Remember that genetics is not necessarily destiny. Random events that occur during development can cause two individuals who are genetically identical to differ in slight ways. For example, the first cloned kitten, CC (short for "carbon copy"), has a completely different coat color pattern than the cat that provided the material for her cloning and has a distinctly different personality than her biological mother.[24] Also, the cloned sheep made by Ian Wilmut at the Roslin Institute of the University of Edinburgh are all genetically identical, but they are of different sizes and have distinct personalities.[25] Therefore parents who are expecting a replacement are probably going to get a child who is not the same as the one who died. Consider their profound disappointment when they discover this and the effect it would have on the cloned child. Such unjust expectations most likely would psychologically damage the cloned child.

Second, it is usually the case that all children are told at some time in their lives that they are special because they are unique; there is no one else like them in the whole world. That is certainly the case with my three children. What do we tell cloned children? Being a carbon copy of another child is not an inspiring start of a healthy self-image. Once again there is the possibility that simply being a clone of someone else may produce psychological damage.

Third, reproductive cloning can hurt the family. Who are the cloned children's parents? They do not really have a proper mother and father who brought them into being as a result of a reproductive choice. What then are they? Cloning erases the biological connection parents normally have to their children and denies the cloned child the place children normally occupy in a family.

Fourth, cloning is an attack on human dignity because cloned children are made to fulfill particular roles. They are made for spare parts or to replace someone else. They are products manufactured to fill roles rather

24. Hays, "Cloned Cat Isn't a Carbon Copy."

25. Wilmut et al., *Second Creation*, 275–80.

than autonomous, self-determining human beings. This empties cloned children of the dignity that all humans should share.

Finally, another potential problem is the specter of eugenics. With cloning comes the ability to pick which embryos should be born. This means that only embryos with particular genotypes will have the privilege of being born. History has shown us that eugenic practices are subject to abuses of power,[26] which makes this potential rather troubling.

These are some of the reasons why I oppose reproductive cloning for humans. It is unsafe, damaging, unwise, and destructive to human dignity.

Michael Buratovich

Notes for Table 20.1

1. Unless otherwise specified, these data are taken from Hill et al., 1787–94.
2. Unless otherwise specified, these data are taken from Rhind et al., "Cloned Lambs," 744–45.
3. Zakhartchenko et al., "Adult Cloning in Cattle," 264–67; Wells et al., "Production of Cloned Calves," 996–1005; Chavatte-Palmer et al., "Calves from Somatic Cells," 1596–1603.
4. De Sousa et al., "Gestational Deficiencies in Cloned Sheep," 23–30.
5. Tanaka et al., "Placentomegaly in Cloned Mouse," 1813–21.
6. Chavatte-Palmer et al., "Calves from Somatic Cells," 1596–1603; Shiga et al., "Calves from Cultured Somatic Cell from Japanese Black Bulls," 527–35.
7. Polejaeva et al., "Cloned Pigs from Adult Somatic Cells," 86–90.
8. Lai et al., "Pigs by Nuclear Transfer Cloning," 1089–92.
9. Cibelli et al., "Cloned Calves from Fetal Fibroblasts," 1256–58.
10. Keefer et al., "Cloned Goats after Nuclear Transfer," 199–203.
11. Ogonuki et al., "Early Death of Cloned Mice," 253–54.
12. McCreath et al., "Gene-Targeted Sheep from Somatic Cells," 1066–69.
13. Chavatte-Palmer et al., "Calves from Somatic Cells," 1596–1603; Kato et al., "Cloning of Calves," 231–37.
14. Denning et al., "Deletion of Genes in Sheep," 559–62.
15. Kishi et al., "Nuclear Transfer in Cattle," 1135–40.
16. Ogonuki et al., "Early Death of Cloned Mice," 253–54; Renard et al., "Somatic Cloning," 1489–91.
17. Zakhartchenko et al., "Cloning in Cattle," 264–67.
18. Ogonuki et al., "Early Death of Cloned Mice," 253–54; De Sousa et al., "Gestational Deficiencies in Cloned Sheep," 23–30.
19. Kato et al., "Cloning Calves," 231–37.
20. Tamashiro et al., "Cloned Mice Have an Obese Phenotype," 262–67; Inui "Obesity," 77–80.

26. Black, *War Against the Weak*, 185–374.

Letter #21

What about Altered Nuclear Transfer/ Oocyte Assisted Reprogramming (ANT/OAR)?

Doc,

I visited Kress and Debbie last week. They reminded me to tell you hello. Kress is done with medical school and has started his internship in anesthesiology.

Kress said that a guy came to Wayne State Medical School and talked about "altered nuclear transfer" as a way to make embryonic stem cells without destroying embryos. He described it as using genetic procedures that ensure that an egg never forms an embryo, even after the egg is subjected to cloning techniques. He showed me his notes from the talk and I didn't understand it very well. The details were a little fuzzy to Kress, but he remembered that it sounded pretty convincing. What do you know about this technique, and can it make embryonic stem cells without destroying embryos?

Matt H.

Dear Matt,

The person who has championed altered nuclear transfer (ANT) and oocyte assisted reprogramming[1] (OAR) as an alternative method of producing embryonic stem cells without destroying embryos is William B. Hurlbut, who was a member of President George W. Bush's Council on Bioethics.

ANT uses the following procedure (fig. 21.1):

A. Remove the nucleus from an unfertilized egg (enucleation).

B. Pretreat an adult cell with reagents that inhibit genes necessary for placental development.

C. Replace the egg nucleus with the nucleus from the adult cell that has been pretreated.

D. Stimulate the embryo with chemicals or a brief electric shock to initiate cell division.

E. Directly grow the cells produced by the embryo in a laboratory culture dish to form embryonic stem cell lines.[2]

The embryo generated with this protocol will make only inner cell mass cells, which produce the embryo proper, and no trophoblast cells, which make the placenta. Therefore, ANT generates entities that can be directly grown in the laboratory to form embryonic stem cell lines.

Entities produced by ANT cannot form the outer layer of trophoblast cells, which form a large proportion of the placenta. According to Hurlbut, the entity made by ANT "lacks the intrinsic potential of an actual organism, but possesses the limited organic powers of a tissue or cell culture."[3] Therefore, the production of embryonic stem cells from eggs subjected to ANT does not, according to Hurlbut, include destroying a human embryo because no embryo existed at any time during the procedure.

1. The term *oocyte* refers to an ovulated egg.
2. Hurlbut et al., "Seeking Consensus," 42–50.
3. Hurlbut, "Altered Nuclear Transfer," 220.

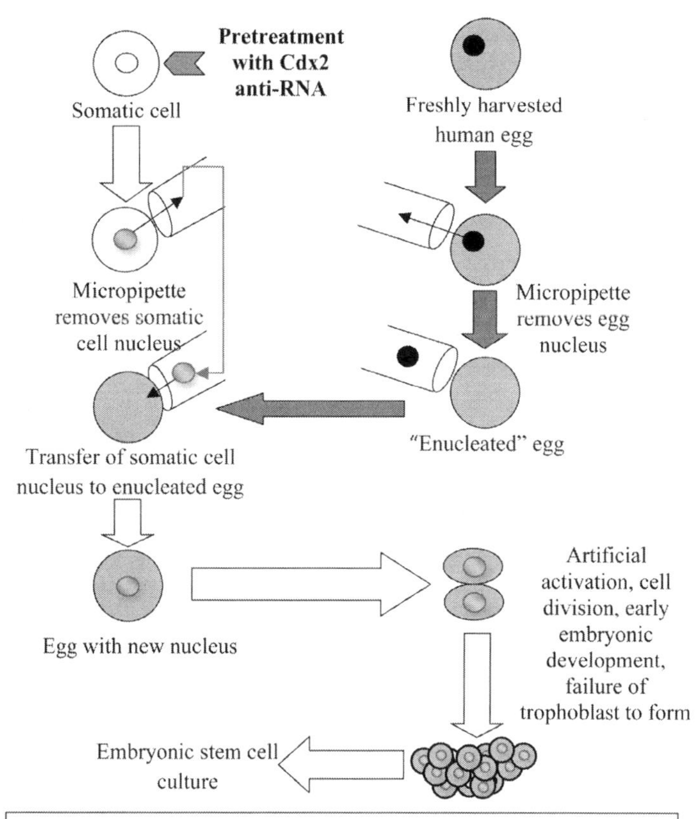

Figure 21.1. The procedure for altered nuclear transfer is shown above. The egg has no potential to form a viable embryo, but only has the ability to form inner cell mass cells, which can be cultured to establish embryonic stem cell lines.

How does someone prevent an early embryo from forming the outer trophoblast cells so that it can be directly cultured to form embryonic stem cells? Hurlbut has proposed inhibiting a gene called *Cdx2* (caudal-type homeobox-2) in unfertilized eggs, prior to replacement of the nucleus. The *Cdx2* gene encodes a protein that is made when the embryo is sixteen to thirty-two cells large and is required for the differentiation of the trophoblast cells,[4] which form the outer layer of the embryo and control the flow

4. Cross et al., "Development of the Placenta," 123–30; Ralston and Rossant, "Stem Cell Origins in the Mouse Embryo," 106–12.

of water and electrolytes into its internal cavity.[5] The trophoblast cells also eventually form the placenta, which attaches the embryo to the inner layer of the mother's womb. Without *Cdx2*, the trophoblast cells neither form properly nor survive, and the embryo collapses into a mass of cells that seems to resemble the inner cell mass (fig. 21.1). *Cdx2*-deficient embryos also fail to implant into the uterus and die rather early during development.[6] Again, because the entities generated by ANT do not possess the potential to form viable embryos, Hurlbut and his colleagues argue that they ought to be regarded as "biologically (and therefore morally) equivalent not to embryos, but to teratomas [i.e., germ cell tumors composed of multiple cell types] and other fragmentary and unorganized growths."[7]

Experiments with mice by Alexander Meissner and Rudolph Jaenisch have demonstrated that ANT is feasible. Meissner and Jaenisch used the protocol described above and successfully produced embryonic stem cells when grown in a laboratory culture dish.[8]

A variation of ANT called oocyte-assisted reprogramming (OAR) forces the expression of a gene called *Nanog* that is only made by the cells of the inner cell mass. The *Nanog* gene is necessary for the inner cell mass cells to maintain their ability to form all adult tissues (pluripotency), and it is never made in the trophoblast cells. Nanog protein is made at low levels in the sixteen-cell stage (morula), and at high levels in the inner cell mass cells once the embryo is five to seven days old (blastocyst stage).[9] Presumably, high levels of Nanog throughout the entire embryo would push all the cells produced by the egg to become inner cell mass cells that could be cultured to form embryonic stem cells (fig. 21.2).

5. Flemming et al., "Cell Adhesion in the Preimplantation Embryo," 1000–1007.

6. Strumpf et al., "*Cdx2* Is Required for Trophectoderm," 2093–2102; Niwa et al., "Trophectoderm Differentiation," 917–29.

7. Hurlbut et al., "Seeking Consensus," 44.

8. Meissner and Jaenisch, "ES Cell from *Cdx2*-Deficient Blastocysts," 212–15.

9. Mitsui et al., "Nanog Required for ES Cells," 631–42; Hatano et al., "*Nanog* Activity," 67–79; Niwa, "Pluripotency," 635–46.

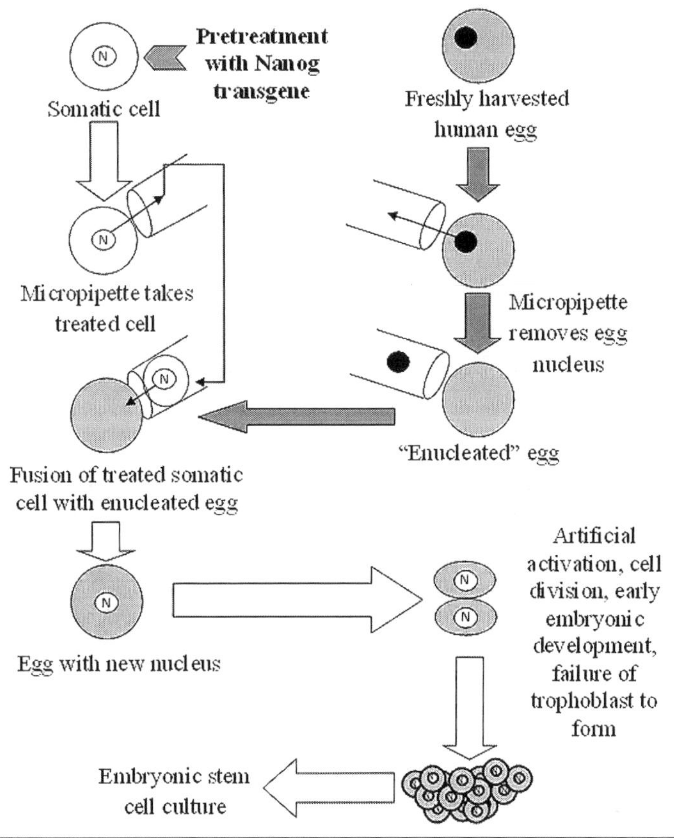

Figure 21.2. The procedure for Oocyte Assisted Reprogramming is shown above. Artificially-generated high levels of the *Nanog* gene would, presumably, push the cells to form only inner cell mass cells and no trophoblast cells. Because the placenta-making cells do not form, no viable embryo forms. The cells can be cultured to establish embryonic stem cell lines.

Presently, no animal experiments have demonstrated the feasibility of OAR.

Altered nuclear transfer and oocyte-assisted reprogramming have garnered the support of several high-level scientists, theologians, and philosophers,[10] but there are several concerns about the efficacy of these procedures. First, with respect to ANT, the *Cdx2* gene is needed for more than just trophectoderm formation. In mice, *Cdx2* is also necessary for several other important events during embryonic development,[11] and is required for the formation of the intestines.[12] Therefore, embryonic stem cells without the *Cdx2* gene "will probably be restricted in their developmental capacity in ways that are impossible to predict but that will probably limit their usefulness in research and clinical applications."[13] William Hurlbut suggests that only a temporary inhibition of *Cdx2* could fix this problem, but this adds a layer of complexity to the procedure and introduces an ethical problem. As McMaster University theologian James C. Peterson has noted, during ANT, the scientist removes something from the embryo that can potentially be returned to it. Since an embryo requires constant protection and feeding for survival and growth, Peterson asks, why would "needing a correction not be part of support?"[14] Thus, according to Peterson, ANT does not really solve the ethical problem surrounding embryonic stem cell derivation. In addition, it is unknown whether human embryos with a functional *Cdx2* gene will die at the same stage as do mice embryos and whether they can be used to make embryonic stem cells. Answering such a question would require research that sacrifices human embryos, and therefore ANT would create the very moral problem it was designed to solve.[15]

Questions also exist about the feasibility of OAR. Several studies have established that making pluripotent stem cells requires more than heightened levels of the Nanog protein.[16] Several labs have established that inserting combinations of different genes into adult cells can transform them into cells that look and behave like embryonic stem cells.[17] Thus, Nanog does not work alone, but must work in combination with other genes to convert cells into pluripotent stem cells. This calls the feasibility of OAR into question.

10. Ethics and Public Policy Center, "Oocyte-Assisted Reprogramming."
11. Chawengsaksophak et al., "*Cdx2* in Development," 7641–45.
12. Traber and Stilberg, "Intestine-Specific Gene Transcription," 275–97.
13. Melton, Daley, and Jennings, "Flawed Proposal," 2791–92.
14. Peterson, "Ethics of ANT," 294–302.
15. Colombo, "Biological and Moral Notes," 645–48.
16. Byrnes, "Flawed Scientific Basis of Altered Nuclear Transfer," 60–65.
17. Yu and Thomson, "Pluripotent Stem Cell Lines," 1987–97.

Finally, ANT and OAR use somatic cell nuclear transfer, which is the same procedure involved in embryo cloning. They require large quantities of human eggs and are expensive and extremely inefficient procedures.[18]

Potential ethical issues also plague ANT/OAR, since these procedures are a form of cloning.[19] Once a cloned embryo is made, a new human life begins. That an embryo might lack enough Cdx2 protein at the right time or too much Nanog protein at the wrong time does not alter the fact that an embryo has been created with a complete human genome and the proper developmental context to become a person.[20] The fact that the embryo made by ANT looks completely normal for the first three or four days of its existence demonstrates this. For these reasons, many bioethicists and scientists "see no basis for concluding that the action of *Cdx2* (or indeed any other gene) represents a transition point at which a human embryo acquires moral status."[21] Thus ANT/OAR, despite the ingenious nature of these proposals and the reverent spirit with which they were offered, remain unsatisfying options.

That's how I see it.

Cheers,

Michael Buratovich

18. Hall et al., "Therapeutic Cloning to Fight Human Disease," 1628–37; Mombaerts, "Therapeutic Cloning," 119–245.

19. Walker, "Philosophical Critique," 649–84.

20. Austriaco, "Critique of a Critique," 172–76.

21. Melton, Daley, and Jennings, "Flawed Proposal," 2792.

Letter #22

Can We Really Use Dead Embryos to Make Embryonic Stem Cells?

Dear DB,

I have talked about this with our basketball chaplain, Papa Joe, and he told me to ask you about this.

I went to church with my parents over Thanksgiving break and the parish priest said something in his announcements about a way to make stem cells from dead embryos. I asked him about it after Mass, and he showed me that announcement. It was some kind of informational piece from his bishop. There was a website on it, which I copied down and found at home. The site described some experiments that allowed the making of embryonic stem cells with dead embryos. Interesting, huh? I have attached the website for you and wonder what you think about this.

Thanks for looking at this.

Richard B.

Dear Richard,

Thanks for the website. There is lots of great stuff there, and I think I will put it on my Senior Seminar syllabus for students to look at.

I am somewhat sympathetic to this proposal despite its problems. The idea is to harvest embryonic stem cells from embryos that show developmental arrest and are, for all intents and purposes, dead. Defining death during the embryonic stages is harder than you think because embryos typically contain a mixture of live and dead cells. How then do you define "dead" under these circumstances?

Donald Landry and Howard Zucker define embryonic death as the inability of the embryo to divide further.[1] "Dead" embryos cease development because the majority of their cells contain genetic abnormalities.[2] Landry and Zucker sharpened their definition of embryo death by examining 444 embryos that were classified by fertility clinics as nonviable. They noted that nonviable embryos had fewer cells and failed to undergo a process called compaction by five days after fertilization.

Glossary

Compaction - At approximately three-to-four days after fertilization, the cells of the early embryo (blastomeres), form a tight, compact, sphere, with tight junctions between them. These tight junctions seal the embryo and isolate the inside cells. The internal cells will become the inner cell mass cells while the external cells will give rise to the trophoblast cells. The compacted, 16/32-cell embryo is called a morula.

Based on these criteria, one-fifth of all embryos classified as nonviable were, in fact, dead.[3]

Because developmentally arrested embryos have some cells that are alive and possess normal numbers of chromosomes, they can be used to make embryonic stem cell cultures.

1. Landry and Zucker, "Embryonic Death and Embryonic Stem Cells," 1184–86.
2. Laverge et al., "Embryos Showing Cleavage Arrest," 425–29; Voullaire et al., "Chromosome Analysis of Blastomeres," 210–17.
3. Landry et al., "Criteria for Embryonic Death," 367–71.

Here's the first problem. Clinical experience has established that sometimes embryos that appear to be abnormal and contain many dead cells can develop normally.[4] Thus, the challenges facing the use of nonviable embryos are, first, establishing that developmentally arrested embryos can make embryonic stem cells, and, second, arriving at a precise definition of a "dead" embryo.

Two major studies have examined the first issue.[5] In both cases, embryos that had stopped growing by three to four days after fertilization only rarely produced embryonic stem cell lines, whereas those that arrested later (five days or more after fertilization) were an excellent source for embryonic stem cells. Thus developmentally arrested embryos can provide a fine source of embryonic stem cells. George Daley from the Harvard Stem Cell Institute noted that in 2005, 134,242 assisted reproductive cycles were reported, and if even a fraction of these embryos are discarded because they show developmental arrest at five days after fertilization or later, "hundreds of thousands of embryos are potentially available for attempts at hES [i.e., human embryonic stem] cell derivation."[6]

With respect to defining "dead embryos," the criteria established by Landry and Zucker draw a distinct line between living and dead embryos. This is a necessary distinction, since sacrificing an early embryo—which is a human person at a very young stage—to make embryonic stem cells is morally repugnant. If fertility clinics rigorously applied these criteria to their embryos left over from in vitro fertilization attempts and donated all dead embryos to research, they could provide researchers with large quantities of embryos for embryonic stem cell research. Any embryonic stem cell line made from dead embryos would not require the killing of living embryos.

Unfortunately, fertility clinics do not presently use Landry and Zucker's criteria for embryo death, but instead use "embryo grading" systems that rank the embryo according to its appearance. Several different embryo grading systems exist, and they produce diverse evaluations of embryos.[7] Studies have also shown that embryo grading by practicing embryologists

4. Green, "Ethically Universal Embryonic Stem-Cell Lines?" 480–85.

5. Zhang et al., "Embryonic Stem Cells from Arrested Embryos," 2669–76; Lerou et al., "Stem Cells from Poor-Quality Embryos," 212–14.

6. Lerou et al., "Stem Cells from Poor-Quality Embryos," 214.

7. Payne et al., "Assisted Reproductive Technique Outcomes," 900–909; Balaban et al., "Comparison of Two Blastocyst Grading Systems," 559–63.

can vary widely.[8] Also, although there is good correlation between the efficiency of embryo implantation and the appearance of the embryo under the microscope, these same studies also show that low-grade embryos can produce successful pregnancies.[9] Just because a fertility clinic might reject an embryo does not mean that it is dead.[10] Thus, even though dead embryos can provide an ethically acceptable source for embryonic stem cell production, the possibility for abuse seems rather large.

Where does this leave us? Making embryonic stem cells from "nonviable" embryos has several advantages, but the rigorous criteria proposed by Landry and Zucker to determine embryonic death best apply only to very early embryos. Unfortunately, these early embryos are poor sources for embryonic stem cells. Somewhat later embryos are far better sources, but accurately determining their status is problematic. While fertility clinics have well-established rules for "grading" embryos, these standards are not consistently applied internationally. Because of this, making embryos from developmentally arrested embryos could involve the killing of human embryos and is, therefore, not attractive. If the Landry/Zucker criteria could be consistently applied and enforced, then this alternative could successfully work, but it seems to me that enforcing these rigid standards is somewhat impossible.

I'm sorry if this bursts your bubble, but that's what I think.

Michael Buratovich

8. Baxter Bendus et al., "Variation in Day 3 Embryo Grading," 1608–15.

9. Gardner et al., "Blastocyst Score Affects Pregnancy Outcome," 1155–58; Balaban et al., "Blastocyst Quality Affects Success," 282–87; Kovacic et al., "Developmental Capacity of Different Types of Blastocysts," 687–94.

10. Balaban et al., "Poor-Quality Embryos Result in Higher Implantation Rates," 514–18.

LETTER #23

Can We Use Embryo Biopsies to Make Embryonic Stem Cells?

Hey Prof,

I went to see a specialist last Tuesday, and she and I had a great time talking about my condition and about science in general. She thinks that any ethical concerns over embryonic stem cells are easily dealt with if we simply use embryo biopsy techniques to make them. I had her explain exactly what she meant by that, and she told me that genetic diagnoses of embryos are done by removing single cells from the embryo and subjecting those single cells to genetic tests. Because this type of analysis has been available for quite some time, it is clear that it does not hurt the embryo. She also said the single cells removed from embryos can be grown in the laboratory to make embryonic stem cells.

The whole thing fascinated me, but I wondered if I heard it right or understood it all. What do you think about taking single cells from embryos to make embryonic stem cells? Can this stop all the fighting over this issue?

Jordan W.

Dear Jordan,

I sure am glad that you were able to see your specialist. You have suffered long enough. It's time to get some answers and get some relief from all that pain you have been enduring ever so bravely.

What do I know about this? Well, something. Your specialist is right; it is possible to remove cells from live embryos and make embryonic stem cell cultures from them. This technique takes advantage of an ability mammalian embryos have to regenerate the loss of a few cells.

The ability of early mammalian embryos to regenerate provides a way to make embryonic stem cells. Single blastomere biopsy (SBB) uses a device called a micropipette (or micromanipulator) to remove single cells (blastomeres) from very young embryos. Such a technique is used for preimplantation genetic diagnosis (PGD), which tests in vitro fertilized embryos for genetic abnormalities prior to transplanting them into a potential mother's womb. Because mammalian embryos undergo extra divisions to replace lost cells, such manipulation does not harm the embryo. More than one thousand children have been born who underwent PGD at the embryonic stage, and the rate of congenital deformities in these children is no higher than that observed in the general population.[1] This suggests that we can subject embryos to SBB without hurting them, although I must caution you that no studies have examined the potential long-term risks of PGD.

Therefore, since human embryos can undergo SBB without harming them, we could potentially use SBB to make embryonic stem cells without destroying human embryos (fig. 23.1).

1. Verlinsky et al., "Preimplantation Genetic Diagnosis," 292–94.

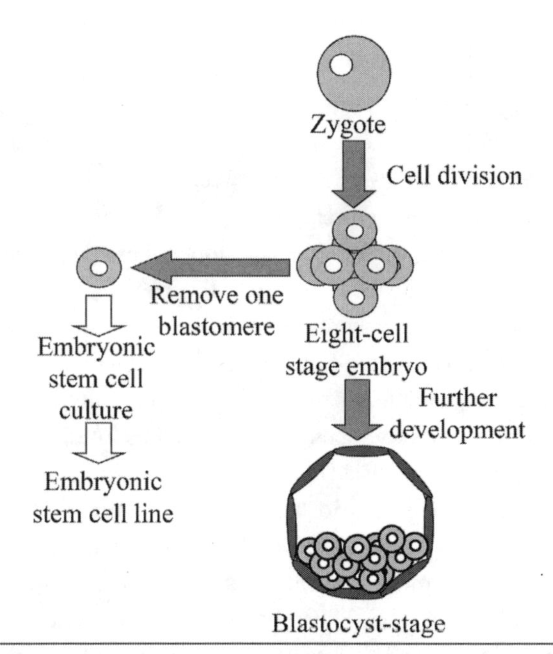

Figure 23.1 The procedure for making embryonic stem cells from eight-cell embryos is shown above. The removal of one cell (blastomere) from an eight-cell stage embryo does not harm the embryo, and this isolated blastomere can be used for genetic analyses. These removed blastomeres can also be cultured to establish an embryonic stem cell line.

Robert Lanza's laboratory at Advanced Cell Technology used SBB on eight-cell stage mouse embryos to isolate cells that were used to produce several pluripotent embryonic stem cell lines without fatally damaging the donor embryos.[2] This demonstrated the feasibility of making embryonic stem cell lines with SBB. The same laboratory also demonstrated that isolated blastomeres from human embryos could produce embryonic stem cell lines, although this particular experiment destroyed the embryos.[3] However, Lanza's research group has since succeeded in making human embryonic stem cell lines from SBB-derived blastomeres without appar-

2. Chung et al., "Stem Cell Lines Derived from Single Mouse Blastomeres," 216–19.

3. Klimanskaya et al., "Human Stem Cell Lines from Single Blastomeres," 481–85; Klimanskaya et al., "Stem Cells from Single Blastomeres," 1963–72; Simpson, "Blastomeres and Stem Cells," 432–35; Chung et al., "Stem Cell Lines without Embryo Destruction," 113–17.

ently harming the embryos.[4] Thus, this alternative might provide a genuine alternative to embryo-destroying techniques that are used to make human embryonic stem cells.

Concerns, however, do exist, since despite the somewhat rosy picture presented earlier, PGD does generate risks to the embryo. Blastomere biopsies of human and mouse embryos cause delays in development and reduce the number of cells in the blastocyst.[5] Time-lapse photography of mouse and human embryonic development after blastomere biopsies reveals particular abnormalities in the developmental behavior of these embryos.[6] Furthermore, not all blastomeres possess an equal capacity to make embryonic stem cells.[7] Thus some embryos might require multiple biopsies to make embryonic stem cell cultures from them, and the risk of harm to the embryo clearly increases as the number of blastomeres removed from it increases.[8]

Additionally, SBB may produce adverse effects many years after birth. Mouse studies indicate that adult mice that developed from embryos that had been subjected to SBB show a higher risk of developing neurological disorders. Since such abnormalities tend to manifest later in life, assessments of SBB that only examine earlier development will gravely underestimate the risks of PGD.[9] Other mouse studies have shown that SBB increases rates of early embryonic and fetal death[10] and alters steroid metabolism in the developing embryo.[11] Such results call into question the purported

4. Chung et al., "Stem Cell Lines without Embryo Destruction," 113–17; Naik, "Scientists Create Stem-Cell Line."

5. Tarin et al., "Human Embryo Biopsy Retards Cleavage," 970–6; Duncan et al., "The Effect of Blastomere Biopsy on Mouse Development," 1462–65.

6. Mouse embryos: Ugajin et al., "Aberrant Behavior of Mouse Embryo Development after Blastomere Biopsy," 2723–28; human embryos: Kirkegaard, Hindkjaer, and Ingerslev, "Human Embryonic Development After Blastomere Removal," 97–105.

7. Lorthongpanich et al., "Single Mouse Blastomeres into Embryonic Stem Cells," 805–13.

8. Cohen, Wells, and Munné, "Removal of 2 Cells from Cleavage Stage Embryos," 496–503; De Vos et al., "Impact of Cleavage-Stage Embryo Biopsy," 2988–96; Goossens et al., "Outcome After the Biopsy of One or Two Blastomeres," 481–92.

9. Yu et al., "Blastomere Biopsy Indicates the Potential High Risk of Neurodegenerative Disorders," 1490–500.

10. Yang et al., "Abnormal Development after Preimplantation Genetic Diagnosis," 1128–33; Sugawara and Ward, "Biopsy of Embryos Affects Development," 234–41.

11. Sugawara et al., "Blastomere Removal Alters Steroid Metabolism," 1–9.

safety of PGD, and without long-term follow-up studies of children who were tested by PGD, we simpy cannot know how safe it is.

However, an even larger ethical problem despoils SBB. Think of it this way Jordan. Let's say your psychology professor thought that you could benefit from a drug that increases intelligence. Thousands of people had taken this drug without ill effects, but nevertheless there were well-documented risks. He is so convinced that you would benefit from this drug that he slips it into your Mountain Dew during lunch, because he wants you to ace the next test. Would you be grateful for his initiative? No, you would be outraged, and completely justified in your outrage. He subjected you to a potentially dangerous substance without your approval. The same can be said for the embryo and SBB. The embryo is a very young human person. It has its own life and we are subjecting it to a potentially lethal procedure without its consent. Saying that parents make such decisions for their children all the time will not wash, because the embryo is not sick; we are subjecting it to SBB without its consent because we want something from it, and we are willing to risk killing it to get it. This is morally objectionable, and concerns like this make SBB a less attractive option for solving the embryonic stem cell debate.

Thanks for asking.

Keep me posted on what the doctor says.

Michael Buratovich

Letter #24

Can We Really Make Embryonic Stem Cells by Putting Genes into Regular Cells?

Dr. Buratovich,

I read in the local newspaper that you can make stem cells by fooling around with a regular cell's genes. This seems like a hoax. Is this for real, and does it stand a chance of working?

Alana C.

Dear Alana,

Tristen told me you were going to ask me this question. I'm glad you did.

This is no hoax, and yes, it stands more than a chance of working. Basically, scientists use genetic engineering to reprogram adult cells to become cells that greatly resemble embryonic stem cells. Such cells are called *induced pluripotent stem cells* (iPSCs).

In 2007, based on work originally done in mouse cells, workers from labs in Japan and the United States used souped-up viruses to transfer particular genes into adult human skin cells and transformed them into iPSCs.[1] Human iPSCs looked like embryonic stem cells, grew like embryonic stem cells, and made all the same genes as embryonic stem cells. They also had the same array of proteins on their surfaces as embryonic stem cells, and, when injected into a laboratory mouse, they were able to differentiate into just as many adult tissues as embryonic stem cells.[2] In comparisons with embryonic stem cells made by destroying embryos, human iPSCs passed every major test.[3] This shows that it is technically possible to use genetic engineering to produce cells that have the properties of embryonic stem cells without the use of embryos.

Can iPSCs potentially be used in a clinical setting? Before such tests are allowed on humans, scientists must show that they can work on animals. To that end, workers from Rudolph Jaenisch's laboratory at Harvard

1. Takahashi et al., "Induction of Pluripotent Stem Cells," 861–72; Yu et al., "Induced Pluripotent Stem Cell Lines," 1917–20. James Thomson's group transformed cells with these four genes: Oct4, Sox2, Nanog, Lin28. This slightly differs from the collection of genes used by Shinya Yamanaka's group, who used Oct3/4, Sox2, Klf4, and c-Myc. Because c-Myc is amplified (extra copies are present) in some human tumors, concerns exist over the use of c-Myc to form iPSCs.

2. iPSCs can differentiate into several cell types with the same efficiency as classical embryonic stem cells—for example, smooth muscle (Xie et al., "Smooth Muscle Induced Pluripotent Stem Cells," 741–48); cardiomyocytes (Zhang et al., "Cardiomyocytes from Induced Pluripotent Stem Cells," e30–e41); fat cells (Taura et al., "Adipogenic Differentiation of Human Induced Pluripotent Stem Cells," 1029–33); nerve tissue (Chambers et al., "Efficient Neural Conversion of iPS Cells," 275–80); hematopoietic and endothelial cells (Cho et al., "Hematopoietic and Endothelial Differentiation of Induced Pluripotent Stem Cells," 559–67).

3. The ends of chromosomes—telomeres—are long in embryonic stem cells. The telomeres in induced pluripotent stem cells are just as long. See Marion et al., "Telomeres in Induced Pluripotent Stem Cells," 141–56. Also, induced pluripotent stem cells make many of the same regulatory microRNAs as embryonic stem cells. See Wilson et al., "MicroRNA Profiling," 749–58.

University made iPSCs from mice that were afflicted with a genetic disease called sickle-cell anemia. Sickle-cell anemia causes deformed red blood cells that get stuck in blood vessels. The clogged blood vessels cause problems in the spleen, kidney, and lungs. People or mice with sickle-cell anemia have poor health overall. Jaenisch and his colleagues used genetic engineering to fix the sickle-cell anemia mutation in these iPSCs. Then they made blood-making stem cells from them and injected them back into the bone marrow of the mice with sickle-cell anemia. After a few weeks, the treated mice were cured of sickle-cell anemia and were normal even twelve weeks after transplantation.[4] Thus, iPSC treatments cured mice with sickle-cell anemia.

In a second experiment, iPSCs were used to cure mice with a genetic disease called hemophilia. iPSCs were differentiated into liver-making cells and infused into the livers of hemophiliac mice, and one week after transplantation, iPSC-treated mice were able to clot their blood and survived cuts, whereas untreated hemophiliac mice bled to death in a few hours.[5] These and other experiments show that it is possible to use iPSCs in a clinical setting to cure genetic diseases.[6] Additionally, several labs have even used adult cells from patients with chronic diseases to make patient-specific iPSCs.[7] Therefore, iPSCs seem close to being ready for clinical trials.

4. Hanna et al., "Treatment of Sickle Cell Anemia Mouse Model with iPS Cells," 1920–23.

5. Xu, Alipio, and Fink, "Phenotypic Correction of Murine Hemophilia A," 808–13.

6. Another research group succeeded in using iPSCs to cure diabetic mice. See Alipio et al., "Reversal of Hyperglycemia in Diabetic Mouse Models," 13426–31.

7. Amylotrophic lateral sclerosis: Dimos et al., "Induced Pluripotent Stem Cells from Patients with ALS," 1218–21; spinal muscular atrophy: Ebert et al., "Induced Pluripotent Stem Cells from Spinal Muscular Atrophy Patient," 277–81; Parkinson's disease: Soldner et al., "Parkinson's Disease Induced Pluripotent Stem Cells," 964–77; Adenosine deaminase deficiency-related severe combined immunodeficiency, Shwachman-Bodian-Diamond syndrome, Gaucher disease, Duchenne and Becker muscular dystrophy, Huntington's disease, juvenile-onset type 1 diabetes mellitus, Down syndrome, and the carrier state of Lesch-Nyhan syndrome: Park et al., "Disease-Specific Induced Pluripotent Stem (iPS) Cells," 877–86.

Glossary

Hemophilia - A rare, inherited bleeding disorder that prevents the blood from clotting after an injury. Hemophilia patients bleed for longer than normal after an injury and this bleeding can damage their tissues and organs, and can even be fatal. Hemophiliacs do not have enough *clotting factors* in their blood. Clotting factors are proteins made by the liver and secreted into the blood. They help clot blood when tissues are damaged.

Since no embryos are made or destroyed in the making of iPSCs, the use and creation of these cells present few ethical problems. However, the way they are produced might compromise their safety. The viruses used to introduce genes into cells can cause infected patients to become very sick,[8] or can introduce deleterious mutations into the cells.[9] Fortunately, this problem is not intractable, since scientists have designed ingenious ways to move the necessary genes into adult cells without using viruses.[10] This shows that there are several ways to bypass the problems associated with using viruses to convert adult cells into iPSCs.

A second concern involves the genes used to convert the adult cell into an iPSC. Greatly increasing the expression of these particular genes in cells can drive them to become tumor-forming cells.[11] Fortunately, some research groups have been able to use different cocktails of genes that are

8. Ailles and Naldini, "HIV-1-Derived Lentivirus Vectors," 31–52; Sourvinos et al., "Retrovirus-Induced Oncogenesis," 226–32; Hacein-Bey-Abina et al., "LMO2-Associated Clonal T Cell Proliferation," 415–19.

9. Trobridge, "Genotoxicity of Retroviral Therapy," 581–93; Okita et al., "Germline-Competent Induced Pluripotent Stem Cells," 313–17.

10. González, Boué, and Belmonte, "Making Induced Pluripotent Stem Cells," 231–42; Papapetrou and Sadelain, "Generation of Transgene-Free Human Induced Pluripotent Stem Cells," 1251–73; Zhou and Ding, "Evolution of Induced Pluripotent Stem Cell Technology," 276–80; Sidhu, "New Approaches for the Generation of Induced Pluripotent Stem Cells," 567–79.

11. Robson et al., "c-Myc and the Pathogenesis of Cancer," 305–26; Vita and Henriksson, "Myc as a Target for Cancer," 318–30.

not as potentially tumor-forming as the original combination.[12] Additionally, several stem cell scientists have established that particular chemicals, in combination with a subset of the genes normally used, can effectively transform various types of adult cells into iPSCs.[13] Some have even used recombinant proteins that pass directly into cells and convert them into iPSCs.[14] Therefore it is now possible to make embryonic stem cells with no viruses, minimal genetic engineering, and no dead embryos. Thus the main safety issues over IPSC production have been largely solved,[15] although some concerns still remain.[16]

One concern is that the production of iPSCs from adult cells seems to induce mutations in cells that were not present before they were converted into stem cells.[17] These might increase the ability of some iPSC lines to cause tumors when implanted into patients.[18] However, since there are ways to test the lines for genetic integrity, this problem may not be insurmountable.[19] Second, these mutations might also induce the expression of surface molecules not normally seen on the surfaces of adult cells, which will activate the immune system to attack and destroy the implanted cells as foreign invaders.[20] Fortunately, this does not seem to occur in the majority of iPSC lines, and some methods for making iPSCs seem to minimize the occurrence of these genetic changes.[21]

12. iPSCs made by virally induced ectopic expression of Oct4/Sox2/Klf4: Aoi et al., "Pluripotent Stem Cells from Liver and Stomach Cells," 699–702; Nakagawa et al., "Induced Pluripotent Stem Cells without Myc," 101–6; iPSC conversion by virally induced ectopic expression of Oct4/Sox2/Nanog/Lin28: Yu et al., "Induced Pluripotent Stem Cells from Somatic Cells," 1917–20.

13. Mali et al., "Butyrate Enhances Derivation of Human Induced Pluripotent Stem Cells," 713–20.

14. Zhou et al., "Generation of Stem Cells Using Recombinant Proteins," 381–84.

15. Sidhu, "New Approaches," 567–79.

16. Pera, "Dark Side of Induced Pluripotency," 46–47.

17. Gore et al., "Somatic Coding Mutations," 63–67; Lister et al., "Hotspots of Aberrant Epigenomic Reprogramming," 68–73; Hussein et al., "Copy Number Variation," 58–62; Martins-Taylor et al., "Recurrent Copy Number Variations," 488–91.

18. Laurent et al., "Dynamic Changes in the Copy Number," 106–18; Amps et al., "Chromosome 20 Minimal Amplicon," 32–44.

19. Tsuji et al., "Safe-Induced Pluripotent Stem Cells for Spinal Cord Injury," 12704–9.

20. Zhao et al., "Immunogenicity of Induced Pluripotent Stem Cells," 212–15.

21. Cheng et al., "Low Incidence of DNA Sequence Variation, 337–44.

Scientists have made iPSCs from cells taken from autopsies[22] and urine.[23] Furthermore, a Finnish research group has successfully made clinical-grade iPSCs that have never been exposed to animal cells or products.[24]

Incidentally, there is an even more exciting find that allows the transformation of cells from one cell type to another without passing through a pluripotent stage. Workers in Doug Melton's lab at Harvard University managed to use genetic engineering techniques to directly reprogram a pancreatic enzyme-making cell into an insulin-making cell.[25] Since then, several mature adult cells have been reprogrammed into distinctly different cell types without passing through an embryonic stage.[26] This technology (known as "reprogramming") might allow regenerative therapy without ever touching embryonic stem cells or any cell like them, but by directly reprogramming a donor cell into another cell.

In conclusion, iPSCs represent an exciting new development that bypasses the needs for eggs and the destruction of embryos, and they are getting closer and closer to clinical trials every day. It is almost certain that they will be used in drug screening within a few years.

I'm glad to help out when I can. Feel free to write again.

Dr. B.

22. Hjelm et al., "Pluripotent Stem Cells from Autopsy-Derived Cells," 219–24.

23. Zhou et al., "Induced Pluripotent Stem Cells from Urine," 1221–28.

24. Rajala et al., "Clinical-Grade Human Embryonic, Induced Pluripotent Stem Cells."

25. Zhou et al., "Reprogramming Pancreatic Exocrine Cells into β-Cells," 627–33.

26. Buratovich, "Direct Conversion of Skin Cells into Neural Precursors," *Beyond the Dish* (blog), February 3, 2012. Online: http://beyondthedish.wordpress.com/2012/02/03/direct-conversion-of-skin-cells-into-neural-precursor-cells.

Letter #25

Mesenchymal Stem Cell Treatments

Dear Mike,

Thanks for coming to the Honors Showcase for our prospective students. These students genuinely appreciate the chance to talk to a professor. Those of us in admissions love it when our faculty members take ownership of the admissions process.

Another thing—my dad had radiation therapy for his cancer and it really wrecked his bone marrow. The doctors told my mom and me about a "mesenchymal stem cell treatment" to restore his bone marrow. Since you're kind of the "stem cell guru" of the school, could you give me some kind of tutorial on mesenchymal stem cells, please?

Thanks again,

Lindsey G.

Dear Lindsey,

Thanks for your note. I love to work with you guys in admissions, and it's great to meet so many fine young men and women and tell them why SAU would be a fine choice for their college experience.

I am sorry to hear about your dad. It's great that they caught his cancer early, but radiation can indeed smash your bone marrow to bits. Stem cell treatments can certainly help him, and I will try my best to introduce you to mesenchymal stem cells.

The term *mesenchymal* comes from the word *mesenchyme*, which refers to embryonic connective tissue. Mesenchymal stem cells (MSCs) refer to stem cells found in loosely woven tissue, but that originate from cells that hang around blood vessel walls.[1] Adult human MSCs were first isolated from bone marrow by Mark Pittenger and his colleagues in 1999,[2] but several other tissues like umbilical cord,[3] adipose tissue (fat),[4] circulating blood,[5] fetal liver,[6] lung,[7] and extracted teeth all house MSCs.[8] In fact, every organ in our body might harbor some MSC population.[9] MSCs can readily differentiate into bone-, fat-, cartilage-, and smooth muscle-making cells, but laboratory work has revealed that they also have the ability to become many different cell types.[10] In the laboratory, MSCs can form liver, lung, and digestive system cells, as well as various types of brain cells[11] and heart muscle cells.[12]

MSCs have tremendous clinical potential. First, they are easily isolated from a small sample of bone marrow and can also be grown in the

1. Crisan et al., "Perivascular Origin," 301–13.
2. Pittenger et al., "Human Mesenchymal Stem Cells," 143–47.
3. Igura et al., "Chorionic Villi of Human Placenta," 543–53.
4. Gronthos et al., "Human Adipose Tissue–Derived Stromal Cells," 54–63.
5. Zvaifler et al., "Mesenchymal Precursor Cells," 477–88.
6. Campagnoli et al., "Fetal Blood, Liver and Bone Marrow," 1198–1201.
7. in 't Anker et al., "Multilineage Differentiation Potential," 845–52.
8. Miura et al., "SHED," 5807–12.
9. da Silva et al., "Post-Natal Organs and Tissues," 2204–13.
10. Buratovich, "Critical Distinctions," *Beyond the Dish* (blog), March 13, 2013, http://beyondthedish.wordpress.com/2013/03/13/mesenchymal-stem-cell-article.
11. Kopen et al., "Marrow Stromal Cells Migrate into Neonatal Mouse Brains," 10711–16; Jiang et al., "Pluripotency," 41–49.
12. Makino et al., "Cardiomyocytes from Marrow Stromal Cells," 697–705; Toma et al., "Cardiomyocyte Phenotype," 93–98; Wang et al., "Mesenchymal Stem Cells Reverses Remodeling," H275–86.

laboratory in larger numbers. Second, MSCs are "invisible" to the immune system.[13] Thus, MSC transplantations between two unrelated people typically do not result in rejection by the immune system, even though the MSCs might lose their invisibility to the immune system under certain conditions.[14] MSCs might even serve to hide other transplanted cells from the immune system. Third, the ability of MSCs to form many different cell types makes them an ideal cell for regenerative treatments.

Bone marrow transplantations help many people with blood disorders, as do treatments with blood cell–making stem cells. Unfortunately, a major problem for some of these transplant patients is graft-versus-host disease (GVHD). GVHD occurs when immune cells present in the newly transplanted donor cells begin to attack the tissues of the transplant recipient. Because MSCs suppress the immune response, they can delay the rejection of skin grafts[15] and possibly quell GVHD. Moreover, MSCs produce chemicals called "cytokines" that support blood formation and potentially enhance regeneration of the bone marrow after damage from treatments like chemotherapy or radiation.[16]

Several scientists have tried to use MSCs in combination with blood cell–making stem cells to lessen GVHD. Safety studies readily showed that MSC transplantations caused no problems.[17] A more detailed study on patients with blood cancers whose bone marrows were depleted and subsequently treated with transplantations of blood cell–making stem cells plus MSCs displayed even more promise. Half of the forty-six patients showed faster bone marrow regeneration without the problem of GVHD. Similar treatments in patients with breast cancer[18] and an inherited disease called metachromatic leukodystrophy[19] have shown comparable success-

13. Shi et al., "Immunomodulatory Properties of Mesenchymal Stem Cells," 1–8.

14. Prigozhina et al., "Stromal Cells Lose Their Immunosuppressive Potential," 1370–76; Huang et al., "Allogeneic Mesenchymal Stem Cells Induce Immunogenicity," 2419–29.

15. Le Blanc and Pittenger, "Mesenchymal Stem Cells," 36–45; Bartholomew et al., "Skin Graft Survival," 42–48; Di Nicola et al., "Stromal Cells Suppress T-Lymphocyte Proliferation," 3838–43.

16. Le Blanc and Pittenger, "Mesenchymal Stem Cells," 36–45.

17. Lazarus et al., "Mesenchymal Progenitor Cells," 557–64.

18. Koç et al., "Rapid Hematopoietic Recovery," 307–16.

19. Metachromatic leukodystrophy is caused by insufficiency of an enzyme called arylsulfatase A. Without arylsulfatase A, nasty chemicals called sulfatides build up in the body and damage organs.

es.[20] Experiments in animals suggest that these combination treatments can also help people with Fanconi anemia (an inherited form of bone marrow failure).[21] Transplantation of MSCs in combination with blood cell–making stem cell transplantation is safe and can possibly reduce side effects and enhance marrow regeneration after cancer treatments.[22] Similarly, MSC transplantations alone can aid patients whose bone marrows were damaged by cancer chemotherapy or radiation and enable them to properly support blood cell–making stem cells.[23]

When administered to the brain after trauma, MSCs tend to protect brain cells from dying and stimulate healing and regeneration.[24] They also have an antioxidant activity that protects nerve cells from particular types of damage.[25]

MSCs might also eventually treat people with inherited bone diseases. Osteogenesis imperfecta (OI) is a genetic disease that causes fragile bones. OI patients have multiple painful fractures, slow bone growth, and progressively deformed bones. Early studies showed that MSC transplantations into human OI patients improved bone structure and mineralization and lessened fractures, but only slightly improved the symptoms of the patients.[26] In a second, larger study, OI children who received MSC transplantations showed faster bone growth rates than those who did not receive MSCs.[27] Studies like these show that regenerative medicine could possibly treat OI and other bone defects.[28]

In animals, MSCs reduce scarring and damage to lungs treated with anticancer drugs. In humans, MSC treatments have greatly helped patients with strains of tuberculosis resistant to several different types of

20. Meuleman et al., "Mesenchymal Stromal Cell Infusion for Metachromatic Leukodystrophy," e11–13.

21. Li et al., "Mesenchymal Stem/Progenitor Cells in *Fancg-/-* Mice," 2342–51.

22. Lazarus et al., "Mesenchymal Stem Cells and Hematopoietic Stem Cells," 389–98.

23. Fouillard et al., "Engraftment of Allogeneic Mesenchymal Stem Cells," 474–76; Lee et al., "Treatment of Leukaemia by Mesenchymal Stem Cells," 1128–31.

24. Longhi et al., "Stem Cell Transplantation for Traumatic Brain Injury," 143–48; Bliss et al., "Cell Transplantation for Stroke," 817–26; Heile et al., "Mesenchymal Stem Cells after Traumatic Brain Damage," 176–81.

25. Lanza et al., "Mesenchymal Stem Cells with Antioxidant Effect," 1674–84.

26. Horwitz et al., "Mesenchymal Cells in Osteogenesis Imperfecta," 309–13.

27. Horwitz et al., "Bone Marrow Transplantation in Children with Osteogenesis Imperfecta," 1227–31.

28. Undale et al., "Mesenchymal Stem Cells for Bone Diseases," 9893–902.

antibiotics.[29] They form cells that line the digestive system and heal damaged intestines and stomachs.[30]

Damaged skin can also benefit from MSCs. Upon recovery from deep burns, burn patients sometimes have too few sweat glands to cool their body through perspiration. MSC transplants into the skin of burn patients regenerated sweat glands and increased the ability of these patients to perspire.[31]

Finally, scientists have exploited the ability of MSCs to suppress the immune system response in order to design treatments for patients whose immune systems attack their own tissues. Physicians call such conditions autoimmune diseases, and they can wreak havoc upon anyone who is unfortunate enough to suffer from one of them. Fortunately, experimental treatments with infusions of MSCs from a patient's own fat have provided relief to patients with certain autoimmune diseases—including rheumatoid arthritis, atopic dermatitis, polymyositis, and multiple sclerosis—who had exhausted all other treatment options.[32] MSC-based treatments for other autoimmune diseases are also being developed.[33]

The abovementioned clinical applications of MSCs represent only a sample of the potential treatments that these wonderfully versatile and useful cells might provide. The following table lists some of the clinical trials that involve MSCs. Both public and private, for-profit institutions recognize the vast utility of these stem cells. Other clinical trials are also in the works.[34]

29. Mezey et al., "Bone Marrow Stromal Cells," 129–35.

30. Okumura et al., "Mesenchymal Progenitor Cell Subset Contribute to the Gastric Epithelium," 1410–22; Ferrand et al., "Bone Marrow–Derived Stem Cells Acquire Epithelial Characteristics."

31. Sheng et al., "Regeneration of Functional Sweat Gland–Like Structures," 427–35.

32. Ra et al., "Stem Cell Treatment for Patients with Autoimmune Disease," 181.

33. Uccelli et al., "Multiple Sclerosis," 649–56.

34. See http://clinicaltrials.gov/ct2/results?term=mesenchymal+stem+cells.

Table 25.1 Clinical Trials Involving Human MSCs

Disease Treated	Type of Study	Location of Trial
Multiple Sclerosis	Phase II	Hospital Clinic de Barcelona and Instituto de Salud Carlos III, Barcelona, Spain
Cerebral Artery Infarction	Phase II	National University of Malaysia
Coronary Artery Disease	Phase I/II	Rigshospitalet, Denmark
Chronic Left Ventricle Dysfunction after a Heart Attack	Phase I/II	University of Miami
Blood Cancers	Phase II	University Hospital of Liege
GVHD	Phase I/II	Samsung Medical Center, Seoul, South Korea
Stroke	Phase II	University Hospital, Grenoble, France
Cirrhosis	Phase I/II	First Affiliated Hospital of Soochow University, China
Refractory Systemic Lupus Erythematosus	Phase I/II	Nanjing Medical University, China
Spinal Cord Injury	Phase I	TCA Cellular Therapy, Louisiana, USA
Poorly Formed Lungs	Phase I	Medipost Co. Ltd., South Korea
Bone Fractures that Fail to Fuse	Phase I	University Hospital of Liege, Belgium
Parkinson's Disease	Phase I/II	Guangzhou General Hospital of Guangzou Military Command, China
Recovery after Knee Surgery	Phase I/II	Osiris Therapeutics, Columbia, MD, USA (ChrondrogenTM)
Arthritis	Phase I	Mesoblast Corporation, Australia
Knee Joint Osteoarthritis	Phase II	Royan Institute, Iran

Disease Treated	Type of Study	Location of Trial
Rheumatoid Arthritis	Phase I/II	Alliancells Bioscience Corporation Limited, China
Type I Diabetes	Phase I/II	Fuzhou General Hospital, China
Heart Attack	Phase I/II	Angioblast Systems, Texas, USA

MSCs show remarkable promise in regenerative medicine, and the clinical potential of these cells has only just begun to be exploited.

I hope this gives you your "education" in mesenchymal stem cells. Please feel free to contact me again if you have more questions.

Cheers,

Michael Buratovich

Letter #26

Cord Blood Freezing—Yes or No?

Doc,

I'm sorry I missed class yesterday, but my stepmom went into labor and I really wanted to be there. I now have a baby sister who is seventeen years younger than me. I feel really old now! My new sister, Tia, is very cute and really beautiful.

Dad paid for her cord blood to be frozen and stored. It was fascinating, but the nurse didn't know that much about it. She said that the umbilical cord blood contains stem cells that can be used to heal her if she gets sick later in life. Dad said that it was expensive, but worth it.

What do you know about this? Did Dad get taken for a high-priced ride, or did he do a good thing?

Andrea B.

Dear Andrea,

Congratulations! I'm sure you will be a great big sister. I look forward to all the pictures that you have taken. I will look for them on Facebook. For yesterday's lecture, see the Blackboard site online.

Did your dad make a wise decision? It depends.

Physicians used to toss umbilical cord blood as medical waste until they discovered that it is a rich source of stem cells. In 1989, Elaine Gluckman performed the first umbilical cord blood transplant from a sibling to a six-year-old brother who had a bone marrow disease.[1] Since then, some twenty thousand umbilical cord blood transplants have been done in children and adults, and cord blood is now being used to treat diseases that were formerly treated with bone marrow.[2] To date, umbilical cord blood transplants have successfully treated leukemias, lymphomas, aplastic anemias, myelodysplasia, hemoglobinopathies, immunodeficiencies, and metabolic storage diseases.[3]

There are some distinct advantages to umbilical cord blood. First of all, it is quite easy, painless, and risk-free to collect, unlike harvesting bone marrow.[4] Second, the immune cells in cord blood do not easily react to foreign substances.[5] This means that tissue matching requirements between cord blood donors and recipients are much less stringent than those used for bone marrow transplants.[6] Therefore, greater numbers of recipients can benefit from a smaller number of donors.

1. Gluckman et al., "Hematopoietic Reconstitution by Umbilical-Cord Blood," 1174–78.

2. Gluckman and Rocha, "Cord Blood Transplantation," 451–54.

3. Liao et al., "Cord Blood," 393–412.

4. Rocha and Locatelli, "Alternative Hematopoietic Stem Cell Donors," 83–93.

5. Garderet et al., "Umbilical Cord Blood," 340–46.

6. Rocha et al., "Comparison of Bone Marrow and Umbilical Cord Blood Transplants," 2962–71; Hwang et al., "Umbilical Cord Blood versus Bone Marrow," 444–53; Shi-Xia et al., "Umbilical Cord Blood and Bone Marrow," 278–84.

Glossary

Leukemias – A cancer of the blood or bone marrow that causes excessive production of abnormal cells that overrun the bone marrow and circulating blood.

Lymphomas – A cancer of lymphocytes. Lymphocytes are a type of white blood cell that plays a role in adaptive immunity.

Aplastic Anemias – A "bone marrow failure" syndrome characterized by a stark deficiency in circulating blood cells and bone marrow cells.

Myelodysplasias – A varied group of bone marrow disorders that prevent the production of normal red blood cells. It is a disorder of bone marrow-based stem cells that causes the production of large quantities of immature cells that fill the bone marrow and bloodstream and prevent tissues from receiving adequate quantities of oxygen.

Hemoglobinopathies – A genetic disorder that causes the production of abnormal hemoglobin that is unable to properly ferry oxygen to cells and tissues.

Immunodeficiencies – A group of diseases that hamstring the immune system and prevent it from working properly. The result is that the immune system is completely unable to fight off diseases.

Metabolic storage diseases – A group of genetic diseases in which the body lacks a particular enzyme that degrades a specific substance. The inability to degrade that substance causes it to accumulate in the body to the point where it kills cells.

That's the good news. Unfortunately, there are some down sides to cord blood banking. For cord blood transplantations to work, the recipient must receive a lot of cells. Because cord blood collections usually only yield a small amount of material, cord blood transplantations work only in children.[7] Mind you, people are working on ways to expand cord blood cells

7. Rocha et al., "Transplants of Umbilical-Cord Blood or Bone Marrow," 2276–85.

before transplantation or even combine cord blood units for transplantation to adolescents and adults. Unfortunately, cord blood expansion is still experimental at this time, and therefore such procedures simply do not exist at this time.[8] However, so-called double umbilical cord transplantation co-infuses tissue-matched cord blood units from two different donors. This procedure was first used in 1999, but since then, in Europe alone, more than one thousand adults have been treated with double umbilical cord transplantation.[9] Unfortunately, double umbilical cord transplantation carries increased risk of tissue rejection. Also, research has shown that cord blood can be stored for at least fifteen years, but the efficacy of cord blood stored for longer periods of time is unknown.[10]

Probably the biggest concern with cord blood banking is this: If your stepsister gets sick, there is probably only a 1/400 to 1/200,000 chance that she will be able to use her own umbilical cord stem cells as a treatment.[11] Why? Think about it. If she has a genetic disease, her cord blood stem cells will possess the same genetic defect. She would benefit from a cord blood transplantation from another donor (allogenic transplantation), but not from a transplantation of her own cord blood (autologous transplantation).[12]

This brings us to the two different types of cord blood banking options: public and private. Private banking is not cheap. It costs about $1000 to $2000 for processing and cryopreservation of the cord blood, plus $100 to $150 per year for cord blood storage. Also, if you bank your cord blood with a private cord blood bank, you have access only to your cord blood. If a sibling has one of these diseases, then the little brother or sister's cord blood can be a gift of life for big brother or sister.[13] Likewise, if the parent has a disease that the child's cord blood might help cure, then again, privately banked cord blood can save the parent's life. In these cases, going with a private cord blood bank is a good idea. What if you do not have this

8. Moise, "Banking Cord Blood Stem Cells," 42–52; Schoemans et al., "Umbilical Cord Blood Transplantation," 83–93.

9. Sideri et al., "Double Umbilical Cord Blood Transplantation," 1213–30.

10. Kobylka, Ivanyi, and Breur-Vriesendorp, "Cryopreserved Cord Blood Cells," 1275–78.

11. Ballen et al., "Cord Blood for Personal Use," 356–63.

12. Johnson, "Autologous Banking," 183–86.

13. Percer, "Umbilical Cord Blood Banking," 217–23.

situation? Then private cord blood banking is probably an expensive option that you most likely do not need.

The other option is public banking, which exists free of charge to those willing to donate. If you donate to a public bank you will not have access to your own cord blood. Also, if too little cord blood is collected, it will be donated to research. Secondly, public cord blood banks screen all donations for genetic abnormalities and infectious diseases, and some 71 percent of all public cord blood donations are rejected.[14] However, cord blood treatments for adolescents and adults will probably come from large cord blood banks, since only these banks will have enough material to treat a grown individual. Also, there are reports that the quality of publicly stored cord blood is superior (i.e., the number of stem cells per unit of liquid is higher) to that of privately stored cord blood MTGN-8659-5643.[15]

Do not get the impression that a public cord blood bank is automatically a government-run entity. There are several cord blood banks that are public but are run by nonprofit institutions that are not affiliated with any government. For example, the Tzu Chi Cord Blood Bank is a nongovernmental, public cord blood bank run by the Buddhist Tzu Chi Stem Cells Center and Buddhist Tzu Chi Medical Center in Hualien, Taiwan. This organization has provided cord blood for almost three thousand cord blood stem cell donations, both domestically and internationally.[16]

Some parents might want to bank the children's cord blood as a kind of insurance, betting that future research will find new ways to use cord blood stem cells. This may be a bet worth making, but it is just that—a bet. If your dad has that kind of faith, then it is his money and he is free to spend it as he sees fit. However, it is a gamble on future research.

That's what I see from where I sit.

Michael Buratovich

14. McCullough and Clay, "Deferral of Potential Umbilical Cord Blood Donors," 124–25.

15. Frangoul and Domm, "Quality of Privately Banked Cord Blood," 1248.

16. Yang et al., "Stem Cell Transplantation in Taiwan and Beyond," 48–51. Also see Bone Marrow Donors Worldwide, which also collects cord blood and is not affiliated with any government: www.bmdw.org.

Letter 26A

Placenta and Umbilical Cord Cells

Dear Dr. B,

Come on—you didn't spill all the beans on that e-mail. Do you think the gamble is worth it or not? What is the potential of cord blood stem cells for future medical treatments? I KNOW you have an opinion on this one. Let's hear it—please.

Andrea B.

Dear Andrea,

Yes, I have an opinion. If you really want to hear it then prepare for a longer e-mail.

Placental and umbilical cord stem cells are some of the most versatile stem cells around. They have multiple uses, but the treatment options offered by these stem cells are steadily expanding as work continues on them.

There are several different types of placental and umbilical cord stem cells. Placental stem cells represent the first type. Everything that passes from the mother to her baby must move through the placenta. This highly specialized structure is located at the inner layer of the uterus and is connected to the baby by means of a series of blood vessels packaged together into a rope-like structure called the umbilical cord. The umbilical cord is connected to the baby's abdomen, and its blood vessels bring nutrition, oxygen, and hormones to the baby's body and return wastes and other materials to the mother. The developing baby is also suspended in a fluid called the amniotic fluid. The amniotic fluid fills a thin membranous sac called the amniotic membrane.

The placenta and amniotic membranes contain at least four main cell types that might be therapeutically useful someday: 1) amniotic epithelial cells, 2) amniotic "mesenchymal" cells, 3) placenta-derived mesenchymal stem cells, and 4) amniotic fluid stem cells. There is some disagreement over the ability of amniotic epithelial cells and amniotic mesenchymal cells to continuously divide and replenish themselves, and therefore many would not call these cells stem cells. Nevertheless, amniotic epithelial and mesenchymal cells can form several different types of cells both in culture and when transplanted into laboratory animals.[1] The capacities of these cells are shown in the table below.

1. Sakuragawa et al., "Human Amnion Mesenchyme Cells," 208–14; Whittle et al., "Human Amnion Epithelial and Mesenchymal Cells," 394–401.

Table 26.1 Placental Cells and Their Derivatives[2]

Human amniotic epithelial cells		Human amniotic mesenchymal stromal cells	
Can differentiate into:		Can differentiate into:	
Fat cells	Nerve cells	Cartilage-making cells	Nerve cells
Cartilage-making cells	Pancreatic cells	Bone cells	Pancreatic cells
Bone cells	Liver cells	Muscle (skeletal)	Blood vessels
Heart cells	Lung cells	Fat cells	Heart cells
	Muscle cells (skeletal)		

Placental mesenchymal stem cells share many features with mesenchymal stem cells from bone marrow such as their spindle shape, adherence to plastic, the presence or absence of several cell surface molecules, and the ability to differentiate into bone, cartilage, or fat cells.[3] However, placental mesenchymal stem cells grow faster than other types of mesenchymal stem cells and also elicit less of an immune response and more powerfully suppress any immune response against them. These properties make placental mesenchymal stem cells therapeutically attractive for diseases that involve the immune system.[4] A biotechnology company, Celgene Cellular Therapeutics, has developed a placental mesenchymal stem cell treatment for particularly intractable cases of an inflammatory bowel disease known as Crohn's disease.[5]

Human amniotic epithelial cells (hAECs) do not grow as well as placental mesenchymal stem cells, but hAECs can differentiate into many different cell types and also share with mesenchymal stem cells the ability to suppress immune responses against them.[6] In laboratory experiments

2. Table derived from Parolini, "Cells from Human Term Placenta," 300–311.

3. Dominici et al., "Minimal Criteria for Defining Multipotent Mesenchymal Stromal Cells," 315–17.

4. Li et al., "Mesenchymal Stem Cells Derived from Human Placenta," 539–47; Kern et al., "Comparative Analysis of Mesenchymal Stem Cells," 1294–1301.

5. National Institutes of Health, "Multi-Center Study to Evaluate the Efficacy of Human Placental-Derived Cells for Crohn's Disease," http://clinicaltrials.gov/ct2/show/NCT01155362?term=PDA001&rank=4.

6. Ilancheran et al., "Stem Cells from Fetal Membranes Display Multilineage Differentiation Potential," 577–88; Bailo et al., "Engraftment Potential of Human Amnion Cells," 1439–48; Li et al., "Immunosuppressive Factors Secreted by Human Amniotic Epithelial Cells," 900–7.

with animals whose lungs have been damaged by chemicals or disease, hAECs can repair damaged lung tissue, preserve lung function, and protect wounded lungs from being harmed further by the immune system.[7] Since hAECs can even differentiate into lung cells and integrate into lung tissue, they might provide a potential treatment for cystic fibrosis. HAECs can also differentiate into insulin-producing pancreatic cells that, when transplanted into diabetic mice, can restore normal blood sugar levels.[8]

Amniotic fluid stem cells were first described in 2007 by Antony Atala and his colleagues at Wake Forest University. Amniotic fluid stem cells have some of the same characteristics as embryonic stem cells, but overall they behave more like somatic stem cells.[9] These cells grow extensively in culture, can form a wide variety of tissues,[10] and show great promise for future treatments. For example, Atala and colleagues have used amniotic fluid stem cells to heal injured lungs[11] and repair damaged kidneys in mice.[12] When grown in three-dimensional scaffolds, AFSCs form good-quality bone tissue that can be successfully transplanted into leg bones.[13] In laboratory animals that have suffered a heart attack, AFSC implantations heal the heart, improve function, and prevent further deterioration.[14] In animals that have suffered a stroke, applications of AFSCs into the brain show improved memory, coordination, and movement. There is also evidence that AFSCs can differentiate into nerve cells and integrate into the brain, which shows that AFSCs may have important clinical applications for the treatment of neurological disorders.[15]

7. Murphy et al., "Human Amnion Epithelial Cells Prevent Lung Injury," 909–23; Murphy and Atala, "Amniotic Fluid and Placental Membranes," 62–68.

8. Wei et al., "Human Amnion Cells Normalize Blood Glucose," 545–52.

9. Carraro et al., "Amniotic Fluid Stem Cells," 2902–11.

10. De Coppi et al., "Amniotic Stem Cell Lines," 100–106.

11. Carraro et al., "Amniotic Fluid Stem Cells," 2902–11; Murphy and Atala, "Amniotic Fluid and Placental Membranes," 62–68.

12. Perin, et al., "Amniotic Fluid Stem Cells in Acute Tubular Necrosis"; Hauser et al., "Stem Cells from Human Amniotic Fluid Contribute to Kidney Recovery," 2011–21.

13. De Coppi et al., "Amniotic Stem Cell Lines," 100–106.

14. Delo et al., "Tracking of Human Amniotic Fluid Stem Cells in the Heart," 1185–94; Bollini et al., "Amniotic Fluid Stem Cells Are Cardioprotective," 1985–94; Lee et al., "Benefits in Myocardial Infarction Using Human Amniotic-Fluid Stem-Cell Bodies," 5558–67.

15. Rehni et al., "Amniotic Fluid Derived Stem Cells Ameliorate Behavioural Deficits in Mice," 95–100; Murphy and Atala, "Amniotic Fluid and Placental Membranes," 62–68.

Human medicine has greatly benefitted from the use of amniotic membranes themselves. Eye surgeons use amniotic membranes for corneal grafts to treat a whole host of ocular conditions, as a patch for several different types of ocular injuries, and as a carrier for growing limbal stem cells in culture for treatments.[16] Amniotic membranes help the limbal stem cell population on the eye surface replenish itself.[17]

Glossary

Limbal stem cells – Stem cells on the surface of the eye. These stem cells reside at the "corneal limbus," which borders the cornea, (the transparent, front part of the eye that covers the iris, pupil and anterior chamber), and the sclera (the opaque, white, fibrous, protective, outer layer of the eye).

The umbilical cord connects the baby to the placenta, and it contains several important stem cells. The outer covering of the umbilical cord, the umbilical cord epithelium, is made by the amniotic membrane. Umbilical cord epithelium contains stem cells that can differentiate into skin cells that will form new skin if grafted on the backs of mice.[18]

Umbilical cord blood from newborns also contains a blood-making (hematopoietic) stem cell that can treat the blood-based disorders I talked about in the last e-mail. Because I have already written about those stem cells, I won't belabor the point here.

Umbilical cord blood also contains cord matrix stem (UCMS) cells, which are found in Wharton's jelly, the jelly-like substance that protects and insulates the umbilical cord blood vessels. Five minutes after birth, in response to temperature changes, Wharton's jelly collapses and clamps down the umbilical cord vessels to stanch bleeding.

Animal experiments with UCMS cells show some surprisingly remarkable possibilities for regenerative therapy. Consider the following results:

16. Dogru and Tsubota, "Ocular Surface Reconstruction," 75–93.

17. Anderson et al., "Amniotic Membrane Transplantation," 567–75; Anderson et al., "Management of Band Keratopathy," 354–61.

18. They are grown in fibroblast-populated collagen gel. See Sanmano et al., "Reconstructed Skin Equivalents," 29–39.

- If human UCMS cells are infused into mice with a defective immune system, they can spread throughout the body and become part of the bone marrow, spleen, teeth, or heart.[19]
- If these cells are transplanted into the brains of rats with a Parkinson-like disorder, the rats' abnormal behavior decreases and their brains grow back many of the neurons they had lost.[20]
- UCMS cells and stem cells derived from amniotic fluid have been used to grow heart valves in culture that can perform adequately under laboratory conditions.[21]
- Infusion of human UCMS cells into mice with a defective immune system did not form any tumors. However, if they are infused into mice with lung tumors, UCMS cells migrate to the tumor and surround it. If UCMS cells are engineered to produce human interferon beta (a compound made by our own bodies that kills cancer cells), the infused UCMS cells shrink the tumor.[22] Combining engineered UCMS cells with the anti-cancer drug 5-fluorouracil shrinks the tumor even more.[23]
- Injecting UCMS cells into the retinas of blind retinas induces the formation of large swaths of photoreceptors.[24]
- By culturing UCMS cells on artificial surfaces in the laboratory, researchers can form joint-specific cartilage that can be used to fix degenerated jaw joints.[25]

19. Erices et al., "Human Cord Blood–Derived Mesenchymal Stem Cells," 555–61; Chang et al., "Mesenchymal Stem Cells from Human Bone Marrow and Umbilical Cord Blood," 679–85.

20. Weiss et al., "Human Umbilical Cord Matrix Stem Cells," 781–92.

21. Amniotic fluid–derived stem cells: Schmidt et al., "Heart Valves Based on Amniotic Fluid Derived Progenitor Cells," I64–70; umbilical cord mesenchymal stem cells: Breymann et al., "Cardiovascular Tissue Engineering," 87–92.

22. Rachakatla et al., "Umbilical Cord Matrix Stem Cell–Based Gene Therapy," 828–35.

23. Rachakatla et al., "Combination Treatment of Metastatic Human Breast Cancer," 662–70.

24. Lund et al., "Umbilical Cord Tissue Rescue Retinal Disease," 602–11.

25. Bailey et al., "Tissue Engineering for Temporomandibular Joint Cartilage," 2003–10.

- When UCMS cells were converted into nerve-like cells, they helped repair damaged spinal cords in rats, especially if they were combined with chemicals that promote nerve regeneration.[26]
- UCMS cells can heal damaged human corneas (the transparent part in the middle of the surface of the eye) in culture.[27] When implanted into the corneas of mice with inherited corneal disorders, UCMS cells improve the structure and function of the cornea.[28]
- Transplantation of umbilical cord blood–derived stem cells into the injured spinal cord of a thirty-seven-year-old woman who had suffered a spinal cord injury improved her sensory perception and movement in her hips and thighs within forty-one days of cell transplantation. Various scans showed regeneration of the spinal cord at the injured site and below it.[29]
- Transfusions of UCMS cells into the livers of thirty Chinese patients with severe liver problems significantly improved liver function and symptoms of liver disease.[30]

Clearly UCMS cells hold exciting possibilities for regenerative therapies.

Cord blood also possesses a pluripotent stem cell called an unrestricted somatic stem cell (USSC).[31] Transplantation of USSCs into pig hearts after a heart attack greatly improves heart function.[32] USSC transplantation into the injured spinal cords of rats promoted shrinkage of the injured site, decreased cell death, increased regrowth of nerve cell extensions (axons), and significantly improved movement.[33] Human umbilical cords also contain umbilical cord perivascular cells (HUCPV) that can serve as a source of bone-making cells.[34]

26. Zhang et al., "Wharton's Jelly Cells Promote Recovery after Rat Spinal Cord Transection," 2030–39.

27. Joyce et al., "Human Umbilical Cord Blood Mesenchymal Stem Cells Heal Damaged Corneal Endothelium," 547–64.

28. Liu et al., "Congenital Corneal Diseases."

29. Kang et al., "Human UC Blood, Improved Mobility," 368–73.

30. Zhang et al., "Liver Function in Cirrhosis Patients," 112–20.

31. Kogler et al., "New Stem Cell from Placental Cord Blood," 123–35.

32. Kim et al., "Preclinical Study of Unrestricted Somatic Stem Cells," I96–104.

33. Schira et al., "Recovery from Acute Spinal Cord Trauma by a Somatic Stem Cell from Human Umbilical Cord Blood," 431–46.

34. Sarugaser et al., "Human Umbilical Cord Perivascular (HUCPV) Cells," 220–29.

Letter 26a

This is a sampling of the cells available in the umbilical cord and placenta. Hopefully, Tia will not need them, but if she does, she might have a great source of stem cells at her disposal.

Michael Buratovich

Letter #27

Can Stem Cells Help People with Spinal Cord Injuries Walk Again?

Hey, Dr. B.,

My mom has a cousin who was in a car accident and has been in a wheelchair for the last two years. He was really active before the accident and this whole thing has been a huge drag for him. He keeps asking me about stem cell treatments for paralytics, but I simply have little to say. He is not a candidate for the embryonic stem cell trial for spinal cord injuries that I read about in the paper. They want people who were "recently injured."

Can you give me something to say to him? I would love to be able to tell him something.

Thanks,

Stephanie L.

Dear Stephanie,

My heart goes out to you and your family. Your cousin must be a very brave chap to take so much loss with such outstanding courage. My hat is off to him and you can tell him that for me.

Spinal cord injuries definitely have received a fair amount of press in the stem cell debate. Consider the words of 2004 Democratic vice presidential candidate John Edwards, who promised that if he and John Kerry were elected, "people like Christopher Reeves are going to walk, get up out of that wheelchair and walk again."[1] This was a ridiculous statement, but it was based on some experiments conducted in 2001 by John Gearhart and Douglas Kerr from Johns Hopkins University. They transplanted embryonic germ cells, which are derived from aborted fetuses (not embryos), into spinal cord–injured rats. The rats with the transplanted stem cells regained some use of their back legs.[2] Videos of this success were placed on YouTube, and this experiment became the talk of the media.[3]

Further examination of the treated rats showed two interesting things. First, the implanted cells only rarely turned into nerve cells. This means that the attempt to replace the dead cells that were destroyed in the accident occurred only rarely. Second, the implanted cells induced recovery by producing chemicals that helped the spinal cord heal. So in the end, it was not clear that the cells did what everyone wanted them to do, which is to replace the dead nerve cells and make the spinal cord as good as new.

Having said all this, is there hope that stem cells can help people with spinal cord injuries? Sure. Here's what we know as of now.

In the nervous system there are two major types of cells. Neurons are the cells responsible for nerve impulses that help us do everything from moving to thinking. Neurons have long extensions called axons that act as extension cords that let them plug into other neurons and communicate with them. Neurons are accompanied by other cells called glial cells. Glial cells nurture, protect, nourish, and support the neurons, and without them neurons die. Think of the glial cells as the neurons' mothers who fuss over them, worry about them, and never stop doing things for them. Also, there are lots of different types of glial cells.

Spinal cord injuries kill neurons and glial cells in the immediate area of the wound and sever the axon connections between neurons far from

1. Krauthammer, "Edwards Outrage."
2. Kerr et al., "Recovery of Rats with Motor Neuron Injury," 5131–40.
3. Kay and Henderson, "Paralyzed Mouse Walks Again," 73–75.

the lesion. Cutting the axon kills it, but the rest of the neuron may survive. Within the injured area of the spinal cord, dead neurons and glia release substantial amounts of chemicals that can kill nearby cells. These events induce extensive inflammation that kills more cells and culminates in a large cyst filled with dead tissue.[4]

If grown in a laboratory dish, damaged neurons can regrow their axons.[5] However, within the damaged spinal cord, neurons can't do this. An injured spinal cord injury contains a brew of proteins and other chemicals that keep neurons from making new axons.[6] Also, glial cells form a "glial scar" in the injured spinal cord to wall off the spinal cord from the immune system and prevent further damage to spinal nerves.[7] Glial scars contain molecules that repel growing axons.[8] Essentially, the damaged spinal cord is converted into a toxic waste dump that neurons don't like, blocked by a thick and high brick wall that axons cannot penetrate. That's why restoration of long-term spinal cord injuries is so challenging.

Believe it or not, the trouble does not end there; the local immune response (inflammation), bleeding, and blood clotting within the spine damage it further. Even cells some distance from the site of injury die off, and one particular group of glial cells called *oligodendrocytes* get knocked around pretty badly. Oligodendrocytes make the insulation for the axons of many spinal nerves. Because this insulation contains a lot of a protein called myelin, the insulation is called the myelin sheath. Myelin sheaths help nerve impulses move much faster through the nerves. Loss of the myelin sheath causes problems with the conduction of nerve impulses,[9] and also makes neurons likely to die.[10] So a spinal cord injury leaves some tissues without any input from the nervous system, and leaves some surviving neurons with axons without insulation. Animal studies have revealed that loss of the myelin sheath continues well beyond a year after a spinal injury.[11]

4. Bunge, "Combination Strategies," 262–69.

5. Bray et al., "Peripheral Nerve Grafts," 5–19.

6. Caroni and Schwab, "Inhibitors of Neurite Growth," 175–79.

7. Okada et al., "Dual Role for Reactive Astrocytes," 829–34.

8. Busch and Silver, "Extracellular Matrix in CNS Regeneration," 120–27; Barnabé-Heider and Frisén, "Stem Cells for Spinal Cord Repair," 16–24.

9. Blight and Young, "Remyelination by Schwann Cells," 15–34; Waxman, "Demyelination," 1–14.

10. Grigoriadis et al., "Axonal Damage in Multiple Sclerosis," 211–17.

11. Totoiu and Keirstead, "Progressive Demyelination," 373–83.

That is the pathology of spinal cord injury. How do we fix it? There are two goals for regenerative treatments of spinal cord injuries: 1) restore the insulating myelin coating that covers many spinal nerves, and 2) make new neurons and glial cells that repopulate the injured spinal cord.

The first goal is somewhat feasible. Cells called "oligodendrocytes" make up the myelin sheath, and if we provide new oligodendrocytes, they should make a new myelin sheath that will fix the problem.[12] Oligodendrocytes form from *oligodendrocyte precursor cells*, also known as OPCs. Several types of stem cells can form OPCs in the laboratory, as can embryonic stem cells.[13] Application of OPCs to laboratory animals with damaged spinal cords replaces lost oligodendrocytes, remakes the lost layer of insulation on spinal nerves, and improves movement.[14] Even though tumor formation is a concern with embryonic stem cell–based treatments,[15] embryonic stem cell–derived OPCs have never formed tumors in laboratory animals.[16]

On January 23, 2009, Geron Corporation announced FDA approval of their clinical study to determine the safety of their embryonic stem cell–produced OPC line GRNOPC1. This cell line was made from an embryonic stem cell line called H1. H1 was one of the original embryonic stem cell lines made by James Thompson and his colleagues at the University of Wisconsin, Madison in 1998. It was also one of the embryonic stem cell lines available when President George W. Bush limited federal funding for embryonic stem cells to experiments that utilized those lines that were already in existence. H1 was one of the embryonic stem cell lines that actress Mary Tyler Moore labeled as "contaminated and therefore useless,"[17] or even "useless or dangerous in treating humans."[18] The doublespeak when it comes to this cell line is clear for all to see.

Geron had treated at least four patients with severe spinal cord injuries, but on November 15, 2011, Geron announced that it was abandoning

12. Eftekharpour et al., "Myelination of Dysmyelinated Spinal Cord Axons," 3416–28.

13. Nistor et al., "Oligodendrocytes," 385–96.

14. Sharp and Keirstead, "Oligodendrocyte Precursors," 434–40; Keirstead et al., "Oligodendrocyte Progenitor Cell Transplants," 4694–705.

15. Howard et al., "Apoptosis-Resistant Embryonic Stem Cells," 37–44.

16. Barnabé-Heider and Frisén, "Spinal Cord Repair," 16–24; Cloutier et al., "Human Embryonic Stem Cell–Derived Oligodendrocyte Progenitors," 469–79.

17. Tyler Moore, "Testimony."

18. "Stem Cell Reversal Fixes Bush's Flawed Compromise."

its embryonic stem cell work and selling off its stem cell division. The company called this a "calculated business move." None of the patients had experienced any severe side effects, but the company simply dropped the whole enterprise to focus on their anticancer drugs.[19]

Other cells, besides embryonic stem cells, can form oligodendrocytes:

a. Stem cells found in the nervous system (neural stem cells or NSCs) from adult spinal cord or fetal brains can form new neurons and oligodendrocytes that re-cover the spinal nerves with myelin and promote recovery of movement to some degree.[20] Unfortunately, NSCs seem to have an inherent disadvantage, since the spinal cord does not provide an environment that encourages neural stem cell differentiation or survival.[21] However, there are some ways to get around this problem, like engineering NSCs to survive,[22] picking the right spot in the spinal cord,[23] or using special organic materials that protect the NSCs.[24]

b. Bone marrow–derived mesenchymal stem cells can form myelin-making cells in culture, and transplantation of these cells into the damaged spinal cords of laboratory animals results in some functional recovery.[25] In humans, a clinical study of thirty-nine spinal cord injured patients who received infusions of bone marrow–based stem cells revealed that after two and a half years the majority of patients showed improved feeling in their legs.[26]

c. Likewise, umbilical cord blood stem cells have the ability to remyelinate spinal cord nerves after injury and promote modest functional

19. Frantz, "Geron Cuts Losses," 12–13.

20. Mothe and Tator, "Transplanted Neural Stem/Progenitor Cells," 176–90; Plemel et al., "Neural Precursors Give Rise to Myelinating Oligodendrocytes," 1891–1910

21. Han et al., "Glial-Restricted Precursor Cells," 1–16.

22. Pan et al., "Granulocyte Colony-Stimulation Factor and Neuronal Stem Cells," 656–64.

23. Yan et al., "Neuronal Differentiation," 318–32.

24. Syková et al., "Polymer Hydrogels," 1113–29; Lesný et al., "Macroporous Hydrogels Based on 2-Hydroxyethyl Methacrylate," 829–33; Hejčl et al., "2-Hydroxyethyl Methacrylate Scaffolds," 67–73; Kim et al., "Scaffolds Seeded with Neural Stem Cells," 841–45.

25. Parr et al., "Mesenchymal Stromal Cells for Central Nervous System Injury," 609–19.

26. Cristante et al., "Stem Cells in the Treatment of Chronic Spinal Cord Injury," 733–48.

recovery.[27] A thirty-seven-year-old woman with a spinal cord injury who was treated with umbilical cord blood stem cells experienced improved sensory perception and movement in her hips and thighs within forty-one days of cell transplantation. Scans showed partial regeneration of her spinal cord at and below the injured site.[28] Also, a nine-year-old boy who had suffered a spinal cord injury as a result of a plane crash was treated with three infusions of umbilical cord stem cell preparations and showed progressive improvements in muscle strength, recovery of sexual, bowel, and urologic function, and a significant decrease in nerve-related pain.[29]

d. Hair follicle stem cells can promote spinal cord repair and myelin sheath restoration in animal spinal cord injuries.[30]

e. Baby and adult teeth contain several different stem cell populations,[31] and one cell type in particular—human adult dental pulp stem cells (DPSCs)—shows a marked ability to repair damaged spinal cords by differentiating into oligodendrocytes. DPSCs also promote spinal cord regeneration by inhibiting molecules in the injured spinal cord that prevent neuron regeneration, and by encouraging nerve cell survival and regeneration.[32]

f. Schwann cells make myelin sheaths for nerves that are outside the central nervous system. Schwann cells can be grown in large quantities in the laboratory from very small nerve biopsies[33] and can be made from bone marrow stem cells,[34] or skin-derived precursor cells

27. Dasari et al., "Axonal Remyelination," 391–410; Matsuse et al., "Peripheral Nerve Regeneration," 973–85.

28. Kang et al., "Case Study," 368–73.

29. Ichim et al., "Stem Cell Therapy of Spinal Cord Injury," 30; Buratovich, "Umbilical Cord Stem Cells form Nervous System–Specific Cells for Spinal Cord Repair Therapies," *Beyond the Dish* (blog), January 22, 2012. Online: http://beyondthedish.wordpress.com/2012/01/22/umbilical-cord-stem-cells-form-nervous-system-specific-cells-for-spinal-cord-repair-therapies.

30. Amoh et al., "Multipotent Hair Follicle Stem Cells," 1865–69; Amoh et al., "Pluripotent Stem (hfPS) Cells," 1016–20.

31. Estrela et al., "Mesenchymal Stem Cells in the Dental Tissues," 91–98; Telles et al., "Pulp Tissue from Primary Teeth," 189–94.

32. Sakai et al., "Human Dental Pulp–Derived Stem Cells Promote Locomotor Recovery," 80–90.

33. Avellana-Adalid et al., "Monkey Schwann Cells," 291–300.

34. Someya et al., "Axonal Regeneration," 600–610.

if coaxed in the laboratory.[35] Secondly, even after extensive growth in culture and storage in liquid nitrogen, Schwann cells still maintain the ability to remyeliate nerves.[36] Schwann cells have a demonstrated ability to remyelinate damaged nerves and promote the survival and regeneration of neurons.[37]

Finally, the spinal cord possesses its own stem cell population. These stem cells divide after spinal cord injury and help form the glial scar. These cells also respond to particular chemicals. Infusion of such chemicals increases the production of neurons from the already existing stem cell population in the spinal cord.[38] Therefore we can stimulate the production of new cells from this group of cells and redirect them to form cells that need to be replaced after a spinal cord injury.

The second goal, the regeneration of lost neurons and glial cells, is harder, but probably more important.[39] The glial scar prevents axons from extending and connecting with other cells. Fortunately, recent small successes suggest that this goal is possible.

The olfactory ensheathing cells in the nasal sinus cavity regenerate throughout a person's lifetime, so harvesting these cells requires minor surgery that does not permanently damage the nose. These special glial cells can remyelinate damaged nerves, promote and assist the growth of axons,[40] and form neurons too.[41] In animal studies, olfactory ensheathing cells repair the structure of the injured spinal cord and restore functional deficiencies, like paw/hand use, coordinated walking, breathing, climbing, and bladder function.[42] Olfactory ensheathing cells seem to greatly

35. Biernaskie et al., "Skin-Derived Precursors," 9545–59.

36. Kohama et al., "Human Schwann Cells," 944–50.

37. Honmou et al., "Exogenous Schwann Cells," 3199–208; Takami et al., "Transplants Improve Hindlimb," 6670–81; Pearse et al., "Functional Recovery," 976–1000.

38. Kojima and Tator, "Ependymal Proliferation," 223–38; Martens et al., "In Vivo Infusions of Exogenous Growth Factors," 1045–57; Xu et al., "bFGF-Responsive Neural Progenitor," 174–79.

39. Watson and Yeung, "What Is the Potential of Oligodendrocyte Progenitor Cells?" 113.

40. Doucette, "Olfactory Ensheathing Cells," 567–84.

41. Huard et al., "Adult Olfactory Epithelium Contains Multipotent Progenitors," 469–86.

42. Barnett and Chang, "Olfactory Ensheathing Cells and CNS Repair," 54–60; recovery of locomotion: Lu et al., "Olfactory Ensheathing Cells Promote Locomotor Recovery," 14–21; recovery of coordinated walking: Ramón-Cueto et al., "Functional

encourage the regrowth of already existing neurons. They are the ultimate "grief counselors" for damaged neurons. A good grief counselor can take damaged, hurt people and help them work through their troubles and circumstances to become whole again. Olfactory ensheathing cells act like neuron grief counselors for spinal neurons with severed axons, and they help the damaged axons pick themselves up, dust themselves off, and function again. There is also evidence that olfactory ensheathing cells can help regenerating neurons cross the glial scar.[43]

Given their tremendous success in animals, olfactory ensheathing cells were implanted into human patients with spinal cord injuries. In Australia, two studies, one conducted a year after surgery and another three years after surgery, showed that olfactory ensheathing cell transplants were safe and caused no significant side effects.[44] A study in Lisbon, Portugal, transplanted olfactory ensheathing cells into the spinal cords of seven paraplegic patients. All but one patient showed significant improvements in sensation, and all seven patients experienced increased movement, with only minor side effects.[45]

In China hundreds of patients with spinal cord injuries have had olfactory ensheathing cell transplants. However, in these surgeries patients did not receive their own olfactory ensheathing cells, but tissue from aborted babies. These patients consistently showed small to moderate improvements in movement and sensation regardless of their age and no significant side effects.[46] While long-term observation of these patients is necessary to draw hard-and-fast conclusions, it appears that olfactory ensheathing cell transplants are safe and potentially beneficial.

Olfactory ensheathing cells are not the only type of stem cell that can help regenerating neurons send their extensions across the glial scar. Bone marrow stem cells, in particular mesenchymal stem cells, seem to favorably

Recovery of Paraplegic Rats," 425–35; recovery of paw use and climbing: Keyvan-Fouladi et al., "Functional Repair of the Corticospinal Tract," 9428–34; recovery of climbing and breathing: Li et al., "Transplantation of Olfactory Ensheathing Cells," 727–31; recovery of bladder function: Ramón-Cueto et al., "Functional Recovery of Paraplegic Rats," 425–35.

43. Richter and Roskams, "Hype or Hope?" 353–67.

44. Féron et al., "Autologous Olfactory Ensheathing Cell Transplantation," 2951–60; MacKay-Sim et al., "Human Paraplegia," 2376–86.

45. Lima et al., "Olfactory Mucosa Autografts," 191–203.

46. Huang et al., "Transplantation of Olfactory Ensheathing Cells," 1488–91; Huang et al., "Olfactory Ensheathing Cell Transplantation," 434–38; Huang et al., "Olfactory Ensheathing Cell Transplantation in Patients with Chronic Spinal Cord Injury," 439–43.

modify the environment of the spinal cord to promote healing. Mesenchymal stem cells suppress the inflammation in the spinal cord that occurs soon after spinal cord injury and does so much damage. They also make a variety of chemicals that stimulate nerve survival, growth, and healing, and can form cell bundles that bridge the glial scar and act as "guiding strands" for regenerating neurons. Thus, even though some doubt the ability of bone marrow stem cells to form neurons, it seems they do make curative contributions to injured spinal cords, and data from clinical trials bear this out.[47]

Further research has revealed that particular molecules can soften the glial scar and induce neurons to grow across it. In these cases, no cells were transplanted; rather, molecules were injected into the injured spinal cords, and neuron growth across the glial scar ensued.[48]

Given the complex series of events that cause the spinal cord to undergo such significant damage, it is unlikely that one strategy alone will suffice to heal a spinal cord injury. Therefore scientists have tested several combination treatments:

- Schwann cells engineered to produce a nerve growth factor called neurotrophin-3;
- Schwann cells engineered to produce neurotrophin-3 and brain-derived neurotrophic factor;
- Schwann cells engineered to make neural adhesion molecules;
- olfactory ensheathing cells with Schwann cells and enzymes that can degrade the glial scar;
- growth factors or oligodendrocytes in combination with olfactory ensheathing cells and enzymes that can minimize the glial scar;
- co-transplantation of Schwann cells and mesenchymal stem cells.[49]

Combining these treatments typically produces more robust improvements in movement and sensation.[50] None of these has been tried in human trials to date, but such treatments represent the next step in spinal cord therapy.

47. Wright, et al., "Bone Marrow Treatment of Spinal Cord Injury," 169–78; Park et al., "Long-Term Results of Spinal Cord Injury Therapy," 1238–47.

48. Hepatocyte growth factor: Jeong et al., "Hepatocyte Growth Factor Reduces Astrocytic Scar Formation and Promotes Axonal Growth Beyond Glial Scars," 312–22; transforming growth factor–α: White et al., "Transforming Growth Factor α Transforms Astrocytes to a Growth-Supportive Phenotype after Spinal Cord Injury," 15173–87.

49. Ban et al., "Combination of Schwann Cells with Mesenchymal Stem Cells," 707–20.

50. Bunge, "Combination Strategies," 262–69.

Finally, induced pluripotent stem cells that were made from the spinal cord injured patient's cells have been differentiated into neural stem cells and transplanted into the spinal cord lesion. These cells show remarkable curative abilities for spinal cord injuries.[51] Even though induced pluripotent stem cells carry some risks of forming tumors, some lines clearly do not form tumors.[52]

So while spinal cord injuries represent an immense challenge for regenerative medicine, the groundwork is set for some future advances. There is a lot more work to do, but there is reason for optimism. Tell your cousin not to give up hope. Human trials are on the horizon, and there is no need to kill embryos for them either.

Sincerely yours,

Michael Buratovich

51. Nori et al., "Human-Induced Pluripotent Stem Cell–Derived Neurospheres," 16825–30; Fujimoto et al., "Transplantation of Human iPS Cell–Derived Long-Term Self-Renewing Neuroepithelial-Like Stem Cells."

52. Tsuji et al., "Safe-Induced Pluripotent Stem Cells," 12704–9.

Letter #28

Bone Marrow–Based Stem Cells

Dear Dr. Buratovich,

My friend back home is in the hospital after having had a bone marrow transplant for his leukemia. He is doing well, but it has been a long haul.

What is so special about bone marrow that allows us to use it to cure people with leukemia? Does bone marrow provide the means to cure us from other diseases as well?

Scott R.

Dear Scott,

Bone marrow and its bounty of different stem cells provide the front-line choice for regenerative medicine. Bone marrow consists of a loose, complex, ribbon-like tissue that is loaded with blood vessels called *stroma*. Stroma is filled with bone-making cells (stromal osteoblasts), bone-destroying cells (osteoclasts), and a whole menagerie of stem cells.

When it comes to stem cells in our bodies, bone marrow hits the jackpot. Bone marrow stem cells include the following:

1. Mesenchymal stem cells (also known as multipotent stromal cells), which can readily form bone, fat, and cartilage, and have the capability to form a host of other cell types;

2. A subpopulation of mesenchymal stem cells called mesodermal progenitor cells, which can form cell types from bones, cartilage, heart, blood vessels, connective tissues, muscles, spleen, glands, membranes that line various cavities (mesothelium), and the tissues of the reproductive and urinary systems;

3. Blood cell–making stem cells called hematopoietic stem cells (HSCs). In adult bone marrow, HSCs are located near the bone surface.[1] Injury or treatment with drugs or radiation that destroys the bone marrow causes the HSCs to move into the peripheral circulation.[2] HSCs are also found in other sites outside the bone marrow, like the spleen;

4. Multipotent adult progenitor cells (MAPCs), which might be another subpopulation of mesenchymal stem cells, can transform into a very wide variety of cell types;

5. Very small embryonic-like stem cells, which are relatively rare and not terribly well understood, but show remarkable potential;[3]

6. Marrow-isolated adult multilineage inducible cells (MIAMIs), which might also represent another subpopulation of mesenchymal stem cells, are found in young or old individuals and possess an ability to form many different cell types;

7. Endothelial progenitor cells (EPCs), which form blood vessels.

1. Murphy, et al., "Myc Function in Stem Cells," 128–37.

2. Wilson and Trumpp, "Bone-Marrow Haematopoietic-Stem-Cell Niches," 93–106.

3. Rataczak, et al., "Pluripotent Embryonic-Like Stem Cells," 153–61; Rataczak, et al., "Stem Cells for Neural Regeneration," 3–12. However some scientists strongly doubt that very small embryonic-like stem cells exist. See Buratovich, "New Paper Throws Doubt on the Existence of Very Small Embryonic Stem Cells in Bone Marrow."

These bone marrow stem cells play central roles in healing, bone formation, and blood cell formation.[4] Furthermore, none of these stem cells functions in isolation since they interact with many other types of cells to execute their functions.

Transplantations of either bone marrow or bone marrow–derived stem cells isolated from circulating blood have been used to benefit, treat, or even cure many different diseases and conditions that range from blood or immune system diseases to metabolic disorders. Bone marrow–based stem cell treatments have also been used extensively in cancer patients to regenerate the bone marrow after aggressive chemotherapy. The table below lists some of the medical conditions that have been treated with stem cells from bone marrow.

Table 28.1 Diseases Treated by Bone Marrow-Derived Stem Cell Transplantations[5]

Cancers	
Brain tumors (medulloblastoma and glioma)	Angioimmunoblastic lymphadenopathy
Retinoblastoma	Multiple myeloma
Ovarian cancer	Myelodysplasia
Merkel cell carcinoma	Breast cancer
Testicular cancer	Neuroblastoma
Lymphoma	Renal cell carcinoma
Non-Hodgkin's lymphoma	Soft-tissue sarcoma
Hodgkin's lymphoma	Ewing's sarcoma
Acute lymphoblastic leukemia	Various solid tumors
Acute myelogenous leukemia	Waldenström's macroglobulinemia
Chronic myelomonocytic leukemia	Hemophagocytic lymphohistiocytosis
	POEMS syndrome
	Myelofibrosis

4. Neiva et al., "Osteoblasts in Regulating Hematopoietic Stem Cell Activity," 1449–54.

5. Shizuru et al., "Hematopoietic Stem and Progenitor Cells: Clinical and Preclinical Regeneration of the Hematolymphoid System," 509–38; Hombach-Klonisch et al., "Adult Stem Cells—Therapeutic Applications," 1301–14; Lodi et al., "Stem Cells in Clinical Practice," 9.

Autoimmune Diseases	
Juvenile diabetes	Behcet's disease
Systemic lupus erythematosis	Rheumatoid arthritis
Sjogren's syndrome	Multiple sclerosis
Myasthenia	Juvenile arthritis
Autoimmune cytopenia	Polychondritis
Scleromyxedema	Systemic vasculitis
Scleroderma	Alopecia universal
Crohn's disease	Buerger's disease
Immunodeficiency Diseases	
Severe combined immunodeficiency syndrome	X-linked hyperimmunoglobulin M syndrome
X-linked lymphoproliferative syndrome	
Blood Disorders	
Sickle-cell anemia	Thalassemia
Sideroblastic anemia	Primary amyloidosis
Aplastic anemia	Diamond blackfan anemia
Pure red cell aplasia	Fanconi's anemia
Amegakaryocytic thrombocytopenia	Chronic Epstein-Barr virus infections
Metabolic Disorders	
Hurler syndrome	Osteopetrosis
Osteogenesis imperfecta	Cerebral X-linked adrenoleukodystrophy
Krabbe leukodystrophy	
Wounds and Injuries	
Limb gangrene	Jawbone replacement
Surface wound healing	
Heart Damage	
Acute heart attack	Chronic coronary artery disease
Liver Disease	
Chronic liver disease	
Liver cirrhosis	

Remember that many of these treatments are still in the experimental stages and that they are usually not cures for these diseases. Nevertheless, patients who suffer from these diseases and have been treated with bone marrow stem cells have experienced at least some relief. Truly bone marrow provides an amazing array of stem cells for regenerative medicine.

I hope this fills in some of the gaps for you.

Michael Buratovich

LETTER #29

Can Stem Cells Heal an Ailing Heart?

Mike,

You might have heard that my Dad is still in serious condition after suffering a bad heart attack.

He was very active and now he seems like a lump of flesh with tubes and monitors. They said that it will be a long haul for him, but that he will probably make it.

I'm not a scientist, but I heard that you gave a talk to a sociology class about stem cells. What are the possibilities for stem cell treatments when it comes to a bad heart? Can Dad possibly benefit now, or in the future, from a stem cell treatment? Please keep it simple.

Sincerely,

Sally I.

Dear Sally,

I was so sorry to hear about your father. I hope your time off was well worth it. Please accept my heartfelt condolences and know that you and he are in our prayers.

With respect to your stem cell question, I think the answer is yes. Consider the story of Bernard van Zyl. At the end of 2000, Bernard, who goes by Bernie, experienced a massive heart attack; in mid-2001 he suffered yet another one. After a series of stents, angioplasties, and many other heart treatments, he still experienced significant chest pain and was diagnosed with congestive heart failure. It was only a matter of time before Bernie's heart failed him for good. He seemed to be a candidate for hospice.

Glossary

Stents – wire metal mesh tubes inserted into a clogged artery to prop it open.

Angioplasty – a procedure that opens a blocked artery by inflating a tiny balloon in it.

Then Bernie found a clinical trial being run by Douglas Losordo at the Caritas St. Elizabeth's Medical Center in Boston, Massachusetts. This study examined the ability of stem cells extracted from bone marrow to heal an ailing heart. Even though he initially received the placebo, he eventually received heart injections of bone marrow stem cells after his health quickly faded. After his treatment he experienced gradual, but undeniable, improvements in stamina and vigor. Bernie became certain that his heart had turned a corner.

Glossary

Placebo – Latin for "I will please." Placebos are substances or procedures a patient accepts as medicine or therapy that have no demonstrated, specific healing activity. Any good effect it might have is probably due to the power of suggestion. The use of placebos in clinical tests is vital in order to establish the healing properties of a substance or procedure.

Today Bernie is a changed man. Instead of being short of breath and suffering from almost constant chest pains, he enjoys an active lifestyle in his seventies. He owes his life to adult stem cell treatments.[1]

The story of Bernard van Zyk is by no means unique. For example, the private company TheraVitae uses a simple technique that harvests circulating stem cells from blood and grows them over a one-week process. The finished product is called Vescell,[2] and administration of this material into the hearts of patients treats advanced heart disease. The heart treatment procedure is performed in Thailand, but the laboratory that does the stem cell isolation and multiplication is located in Israel. Once blood is collected from the patient, it is sent to Israel where it is processed, then shipped to Thailand for administration.[3] Osiris Therapeutics, Inc., is another company that offers similar stem cell treatments.[4]

Bob Grinstead, a seventy-year-old retired computer salesman, went to Bangkok, Thailand, in March 2005 for a Vescell treatment. He had suffered a massive heart attack in 1990, and since then had had two coronary artery bypass graft surgeries and twenty-four angioplasties. He was so weak that even washing himself wore him out. His doctors told him that he had exhausted standard medical treatments. While conducting a search via the Internet, Bob found TheraVitae. After undergoing the treatment in Bangkok, Bob's heart went from being weak to working again. He can go for

1. van Zyl, *Adult Stem Cell Therapies Saved my Life*, 7.

2. See http://archive.org/details/VesCellAdultStemCellTherapyforHeartDiseasebyVescell.

3. Kalman, "Stem-Cell Prospect."

4. See http://www.osiristx.com.

short walks and has even taken several trips. As Bob himself says, "I feel like living life again."[5] Several other stories from former heart patients who have received and been aided by similar treatments are also available.[6]

How might stem cells fix a damaged heart? That's an involved question that requires kind of a long answer. First, let's consider the heart. The heart has four chambers and is divided into a left and right side, each with an upper and a lower chamber. The lower chambers are larger and more powerful than the upper ones. The right side receives the blood from our body and runs it to the lungs, where the red blood cells load up with the oxygen we take in with every breath and get rid of the carbon dioxide we breathe out. The left side retrieves the blood from the lungs and sends it to the rest of the body. Since our blood ferries everything through our bodies, from oxygen to food to hormones, the heart must keep blood continuously moving. When the heart stops pumping, a person dies.

Heart muscle needs oxygen just like the rest of our body. Unfortunately, the heart is supplied by only a few small blood vessels that must service the entire heart muscle. If these small blood vessels get clogged with cholesterol or blood clots, they narrow. Narrow blood vessels mean that less oxygen gets to the heart muscle. A heart attack occurs when a small heart vessel gets so clogged that the heart muscle suffocates and begins to die.[7]

A large section of dead heart cells causes the formation of a scar. Scar tissue, however, does not contract, and a heart attack can greatly decrease the efficiency of the heart and compromise its ability to send blood throughout the rest of the body. The heart's troubles, however, do not end there. The scarred area is not able to withstand the constant wear and tear associated with continuous contraction. Thus the remaining heart muscle has to work harder in order to compensate for the dead heart muscle. When overworked, heart muscles react the same way a person's arms do if he or she is required to constantly lift heavy weights: they enlarge. This is called "remodeling." Although large muscles look good in arms and legs, an enlarged heart muscle is disastrous. The heart needs to contract and relax in a cyclical and highly coordinated fashion, and enlarged heart muscle has trouble doing that. For this reason, a heart attack can cause someone's heart to deteriorate further until it fails to work altogether.[8] Heart disease

5. Garger, "Take Heart."

6. Do No Harm—The Coalition of Americans for Research Ethics. See http://www.stemcellresearch.org.

7. Fox, *Human Physiology*, 389–471.

8. Cohn et al., "Cardiac Remodeling," 569–82; Opie et al., "Controversies in

is the leading cause of death worldwide. In the United States alone, there are approximately 1.1 million heart attacks per year.[9]

Treatments for heart failure usually include medications that lower blood pressure, protect the heart from overstimulation by the nervous system, expand the blood vessels around the heart, or fix a fitful heart rhythm.[10] However, even though the use of these drugs can increase one's quality of life and even prolong it, they do not address the problems that result from a heart attack. To truly fix a heart that has suffered a heart attack, it is necessary to replace the scar with thriving heart muscle that contracts and does not tax the rest of the heart.[11] Fortunately, regenerative cell therapy is a possibility for the heart. Several different types of stem cells have the ability to find the injured heart and improve it.[12] Such work has been done in animals and human patients. The question is, which cells are the best for fixing the heart?

Heart-Fixing Stem Cells

Muscle Satellite Cells

The first stem cells that made it to human clinical trials for heart patients came from skeletal muscles. Skeletal muscles are usually attached to tendons and are responsible for movements of the skeleton. These muscles contain stem cells (muscle satellite cells) that can divide and become mature, functioning muscle in response to muscle injury or bodybuilding.[13] About ten years of work in animals have shown that small muscle biopsies can be used to grow muscle cells in the laboratory. When transplanted into the hearts of sick animals, cultured muscle cells make their home there and seem to improve heart function.[14]

Ventricular Remodeling," 356–67.

9. Orlic and Cannon, "Hematopoietic Stem Cells for Myocardial Regeneration," 9–27.

10. Jessup and Brozena, "Guidelines for the Management of Heart Failure," 497–506.

11. Adler et al., "Cardiomyocyte Lineage," 1073–81.

12. Passier et al., "Stem-Cell-Based Therapy," 322–29; Davis and Stewart, "Cardiac Repair," 489–508.

13. Gates et al., "Muscle-Derived Stem Cells," 68–76.

14. Dowell et al., "Stem Cell Transplantation in the Heart," 336–50. For improvement in animal models of chronic heart failure, see He et al., "Autologous Skeletal Myoblast Transplantation," 1940–49. For improvement in animal models of ischemia-reperfusion,

These findings in lab animals paved the way for clinical trials on human patients who had suffered heart attacks and were treated with their own skeletal muscle stem cells. Early, small studies showed that these treatments were easy to do, relatively safe, and improved the heart function and symptoms of heart attack patients.[15] Unfortunately, there were some caveats to this good news. Even though skeletal muscle cells can integrate into the damaged heart, they do not make proper contacts with heart muscle cells and remain, for the most part, functionally isolated from the rest of the heart muscle.[16] This caused the hearts that had received skeletal muscle implants to beat abnormally fast.[17]

Larger clinical studies, like the MAGIC and SEISMIC trials, failed to show significant improvements between the treated group and the placebo groups after six months.[18] However, in these studies abnormal heart rhythms were not a significant problem for those patients who had received skeletal muscle treatments, and treated patients in the SEISMIC trial showed significantly fewer symptoms.[19] Additionally, skeletal muscle treatments seem to prevent, or at least delay, cardiac remodeling. In response to these data, the Florida-based company Bioheart is seeking European marketing approval for its MyoCell technology, which uses skeletal muscles to treat acute heart attack patients, although some heart specialists think that the implementation of this technology is premature at this time.[20]

Skeletal muscle cells might provide even more possibilities for heart regeneration. There is a group of skeletal muscle stem cells that can form heart muscle.[21] Techniques also exist to purify these *spoc* (skeletal-based precursors of cardiomyocytes) cells from the rest of the muscle cells.[22]

---

see Jain et al., "Cell Therapy Attenuates Ventricular Remodeling," 1920–27.

15. Smits et al., "Catheter-Based Intramyocardial Injection," 2063–69; Menasché et al., "Autologous Skeletal Myoblast Transplantation," 1078–83; Siminiak et al., "Phase I Clinical Study," 531–37; Siminiak et al., "POZNAN Trial," 1188–95.

16. Léobon et al., "Myoblasts Are Functionally Isolated," 7808–11.

17. Veltman et al., "Four-Year Follow-Up," 1386–96.

18. MAGIC trial: Menasché et al., "Myoblast Autologous Grafting in Ischemic Cardiomyopathy (MAGIC) Trial," 1189–1200; SEISMIC trial: Duckers et al., "SEISMIC Trial," 805–12.

19. Sherman, "MyoCell Therapy and the SEISMIC Trial," 1–11.

20. Haider et al., "MyoCell," 611–21.

21. Winitsky et al., "Muscle Cells Can Differentiate into Beating Cardiomyocytes," 662–71.

22. Nomura et al., "Skeletal Muscle for Cardiovascular Regeneration," 293–300.

Transplantation of purified spocs into the hearts of laboratory animals that have had heart attacks produces significant improvement in heart function and structure.[23] A similar cell type exists in human muscle. When these human spocs are isolated and transplanted into mice that have experienced heart attacks, they also greatly improve heart structure and function.[24] This promising approach for heart attack treatment awaits human clinical trials.

Other possibilities include genetic engineering of skeletal muscle cells. Introducing genes into skeletal muscle cells that help them properly connect with heart muscle cells eliminates the risk of abnormal heart rates.[25] Also, engineering skeletal muscle cells with genes that encode specific signaling proteins transforms them into heart muscle cells.[26] This technology, coupled with the possibility of isolating and culturing skeletal muscle stem cells, might provide clinicians with a large source of heart muscle cells in the future.

Bone Marrow Stem Cells

Whole Bone Marrow

The next possible group of heart healers includes our own bone marrow stem cells. Private companies have exploited bone marrow stem cells to treat heart patients, and for good reason. Bone marrow contains a wide variety of cells that can become many different types of cells. At least three types of stem cells are found in the bone marrow:[27]

1. hematopoietic stem cells that make all the blood cells;
2. mesenchymal stem cells that form fat, muscle, bone, cartilage, and a host of other cell types;
3. endothelial progenitor cells (EPCs), which form blood vessels.[28]

23. Oshima et al., "Differential Myocardial Infarct Repair," 1130–41.

24. Okada et al., "Myogenic Endothelial Cells," 1869–80.

25. Roell et al., "Connexin 43-Expressing Cells," 819–24.

26. Xiang et al., "Wnt11 Enhances Cardiomyogenic Differentiation," 790–96.

27. Lee et al., "Bone Marrow–Derived Multipotent Stem Cells," 1602–3.

28. For hematopoietic stem cells, see Huang et al., "Hematopoietic Stem Cells," 1851–59. For mesenchymal stem cells, see Phinney and Prockop, "Mesenchymal Stem/Multipotent Stromal Cells," 2896–902. For endothelial precursor cells, see Young et al., "Endothelial Progenitor Cells," 421–29.

More exist, but less is known about them at this time.[29] Physicians and technicians can extract these three cell types and use them to help regenerate sick tissues. Some bone marrow stem cells can also be grown in the laboratory to make more cells if needed.[30]

When transplanted into the hearts of laboratory animals that have had heart attacks, whole bone marrow improves heart function and prevents heart remodeling.[31] There was, however, no consensus on how bone marrow cells helped the heart, since experiments from several different laboratories produced conflicting results.[32] Nevertheless, in all cases bone marrow implantations helped damaged hearts.[33]

Despite the uncertainties regarding *how* bone marrow helps damaged hearts, over twenty clinical trials have been conducted that used bone marrow extracts to treat over nine hundred patients who had recently suffered heart attacks.[34] These studies have firmly established that bone marrow heart transplantations are safe and helpful for human patients.[35] Long-term follow-up studies have shown that some benefits remain after two[36] and four years.[37]

29. Ratajczak et al., "Bone Marrow–Derived Stem Cells," 307–19.

30. McNiece, "Clinical Applications of Hematopoietic Cells," 839–45.

31. Orlic et al., "Bone Marrow Cells Regenerate Myocardium," 701–6; Orlic et al., "Mobilized Bone Marrow Cells," 10344–49.

32. Balsam et al., "Haematopoietic Fates in Ischemic Myocardium," 668–73; Murry et al., "Haematopoietic Stem Cells Do Not Transdifferentiate," 664–68.

33. Bone marrow cells help hearts by 1) making new blood vessels that feed the heart muscle (Yoon et al., "Bone Marrow Regenerate Myocardium," 326–38); 2) secreting molecules that help heart muscle survive and flourish (Fazel et al., "Balance of Angiogenic Cytokines," 1865–77); 3) fusing with heart muscle cells (Nygren et al., "Cell Fusion, but Not Transdifferentiation," 494–501); 4) differentiating into heart muscle cells that replace the dead ones (Rota et al., "Bone Marrow Cells Adopt Cardiomyogenic Fate," 17783–88); and 5) stimulating the growth and differentiation of endogenous stem cells in the heart to repair the heart (Hatzistergos et al., "Mesenchymal Stem Cells Stimulate Cardiac Stem Cells," 913–22; Loffredo et al., "Bone Marrow–Derived Cell Therapy Stimulates Endogenous Cardiomyocyte Progenitors," 389–98).

34. Davis and Stewart, "Autologous Cell Therapy for Cardiac Therapy," 489–508.

35. For a summary of these clinical trials and the metanalyses of them, see Buratovich, "Whole Bone Marrow Transplantations into the Heart: Hope or Hype?" *Beyond the Dish* (blog), August 5, 2011. Online: http://beyondthedish.wordpress.com/2011/08/05/whole-bone-marrow-transplantations-into-the-heart-hope-or-hype/.

36. Assmus et al., "Bone Marrow–Derived Progenitor Cells in Acute Myocardial Infarction," 89–96.

37. Cao et al., "Long-Term Myocardial Functional Improvement," 1986–94;

These studies examined the ability of bone marrow transplants to treat patients who had recently suffered from heart attacks. However, can bone marrow cells help the patient who has lived with a poorly functioning heart for a while? Make that another yes. Several clinical studies have examined efforts to patch up chronically damaged hearts with bone marrow stem cells either in combination with or independent of coronary artery bypass graft surgery. The combined data of fifteen studies conducted over seven years have shown that, for patients with advanced heart disease, bone marrow stem cell transplants have once again proven to be safe and relatively effective. In most cases, stem cell transplantations improved heart function and reduced those bodily signals that indicate that the heart is failing. In the case of chronic heart disease, one metanalysis also showed an association between overall improvement of the heart and bone marrow treatments.[38] Bone marrow stem cell treatments, therefore, can help patients who have struggled with a "bad ticker" get a renewed heart that works better.

Mesenchymal Stem Cells

The previously mentioned studies used whole, unfractionated bone marrow, but several groups have also studied the ability of specific stem cells purified from whole bone marrow to treat damaged hearts. Mesenchymal stem cells (MSCs),[39] for example, which do not play a direct role in blood formation, show an astounding ability to form different cell types.[40] When it comes to treating sick people, MSCs have a vital advantage over other somatic stem cells: they are inherently invisible to the immune system.[41]

Dohmann et al., "Postmortem Findings," 521–26. Postmortem analysis of hearts from patients who participated in a clinical trial showed that the bone marrow cells differentiated into blood vessels and heart muscle.

38. Abdel-Latif et al., "Systematic Review and Meta-Analysis," 989–97; Jiang et al., "Meta-Analysis," 667–80.

39. Liechty et al., "Human Mesenchymal Stem Cells Engraft," 1282–86; Devine et al., "Mesenchymal Stem Cells Distribute to a Wider Range of Tissues," 2999–3001.

40. Muscle: Ferrari et al., "Muscle Regeneration," 1528–30; cardiomyocytes: Makino et al., "Cardiomyocytes from Marrow Stromal Cells," 697–705; bone and cartilage: Prockop, "Stem Cells for Nonhematopoietic Tissues," 71–74; Ashton et al., "Bone and Cartilage by Marrow Stromal Cells," 294–307; fat: Shen et al., "Fat Cells," 117–24; tendon: Young et al., "Mesenchymal Stem Cells for Achilles Tendon Repair," 406–13; Crovace et al., "Cell Therapy for Tendon Repair," 281–83.

41. Pittenger et al., "Multilineage Potential," 143–47; Devine et al., "Stealth and Suppression," S76–82; Majumdar et al., "Surface Molecules on Human Mesenchymal Stem Cells," 228–41.

MSCs transplanted from one person to another person are usually not rejected.[42] Therefore donated MSCs can potentially provide healing material not only for various diseases, but also for a wide variety of people.

MSCs are also widely distributed throughout the body in places other than bone marrow. Since they are found in fat,[43] umbilical cord blood,[44] placenta,[45] amniotic fluid,[46] circulating blood,[47] fetal liver,[48] lung,[49] and pulled baby teeth,[50] they are readily available and easily extracted from patients for clinical use.

Transplantation of MSCs into the hearts of mice,[51] rats,[52] and pigs[53] that have just experienced heart attacks improves heart function, and these implanted MSCs differentiate into heart muscle cells, blood vessels, and blood vessel-specific smooth muscles,[54] albeit at a low rate.

42. McIntosh and Bartholomew, "Stromal Cell Modulation of the Immune System," 324–28; Liechty et al., "Human Mesenchyme Stem Cells Engraft," 1282–86; Bartholomew et al., "Baboon Mesenchymal Cells," 1527–41.

43. Hematti, "Solid Organ Transplantation," 262–73.

44. Bieback and Kluter, "Mesenchymal Stromal Cells from Umbilical Cord Blood," 310–23.

45. Parolini, et al., "Cells from Human Term Placenta," 300–311.

46. Marcus and Woodbury, "Do Not Discard," 730–42.

47. Roufosse et al., "Circulating Mesenchymal Stem Cells," 585–97.

48. O'Donoghue and Chan, "Human Fetal Mesenchymal Stem Cells," 371–86.

49. Murphy et al., "Stem Cells in Airway Smooth Muscle," 11–14.

50. Miura et al., "SHED," 5807–12.

51. Grauss et al., "Mesenchymal Stem Cells Improve Left Ventricular Function," H2438–47.

52. Tang et al., "Mesenchymal Stem Cells Participate in Angiogenesis," 353–61; Yoon et al., "Pre-Treated Mesenchymal Stem Cells," 277–84; Hou et al., "Transplantation of Mesenchymal Stem Cells," 220–28.

53. Mesenchymal stem cells administered thirty minutes after a heart attack can prevent a pig heart from deteriorating. See Makkar et al., "Allogenic Bone Marrow–Derived Mesenchymal Stem Cells," 225–33. If given two weeks after the heart attack, mesenchymal cells can significantly improve the function and appearance of the infarcted heart. Microscopic examination also shows that the implanted mesenchymal stem cells became cardiomyocytes. See Shake et al., "Mesenchymal Stem Cell Implantation," 1919–26; Okura et al., "Cardiomyoblast-Like Cells from Human Adipose Tissue-Derived Mesenchymal Stem Cells," 417–25.

54. Toma et al., "Cardiomyocyte Phenotype," 93–98; Quevedo et al., "Trilineage Differentiating Capacity," 14022–27.

Glossary

Human drug trials are conducted in phases. These phases are:

Preclinical studies – Test tube and animal studies to determine if drug shows clinical promise.

Phase 0 – First trials conducted with-human at low dosages to determine safety.

Phase 1 – Small groups of individuals are given the drug at different dosages to assess safety.

Phase 2 – Larger groups of sick people are given the drug or a placebo to determine if the drug treats the ailment it was designed to treat at a safe dosage.

Phase 3 – The effectiveness of the study drug or procedure being examined is compared with another drug whose effectiveness is established. These studies employ large groups of people over long periods of time.

Phase 4 – Post-Marketing Surveillance. Once the drug has been approved, its use in patients is monitored for side effects and effectiveness of treatment.

Several clinical trials in human patients have demonstrated that MSCs are not only safe but that they help the hearts of those who have just experienced a heart attack.[55] MSCs have also been shown to improve the condition of patients with advanced heart disease.[56] Presently, over thirty different ongoing Phase 1 and Phase 2 clinical trials are further evaluating the safety and ability of MSCs to treat patients who have recently experienced a heart attack, who suffer from congestive heart failure, who are undergoing heart surgery, or who suffer from heart-related chest pains (angina pectoris).[57]

55. Katritsis et al., "Transcoronary Transplantation," 321–29; Hare et al., "Randomized, Double-Blind, Placebo-Controlled, Dose-Escalation Study," 2277–86; Hoogduijn et al., "Donor-Derived Mesenchymal Stem Cells," 222–30; Yang et al., "Delivery Via a Noninfarct-Relative Artery," 380–85; Williams et al., "Functional Recovery and Reverse Remodeling," 792–96.

56. Chen et al., "Intracoronary Transplantation," 552–56; Mohyeddin-Bonab et al., "Old Myocardial Infarction," 467–73.

57. There were thirty-two trials as of April 15, 2012. See http://clinicaltrials.gov/ct2/results?term=mesenchymal+stem+cells%2C+heart.

Once implanted into a heart that has recently suffered a heart attack, MSCs enter a very hostile environment, and consequently the majority of implanted MSCs tend to die. Therefore, some enterprising scientists have tried pretreating MSCs with stimulating molecules, preconditioning them in culture, or genetically engineering MSCs with genes that make them live longer. Transplantation of these modified MSCs into the hearts of animals that have experienced recent hearts attacks have yielded even more impressive results.[58] By using genes that force mesenchymal stem cells to form heart muscle, even greater numbers of implanted cells turn into heart muscle, which also improves heart healing.[59] While these results are very promising, they have occurred only in animals. So far, no human clinical trials with genetically engineered mesenchymal stem cells have been completed.

An alternative strategy to increase the survival of transplanted MSCs involves "preconditioning" them before transplantation. MSC preconditioning consists of growing MSCs in low oxygen concentrations or in the presence of harmful chemicals. Under such conditions, the MSCs adapt to the harsh culture environment and toughen up so that they survive better after transplantation into the heart. Animal studies with preconditioned MSCs demonstrate that these cells survive significantly better in the heart after a heart attack than non-preconditioned MSCs and also do a better job improving heart function after a heart attack than their non-preconditioned counterparts.[60]

Endothelial Progenitor Cells

Another specific bone marrow–derived stem cell that shows tremendous promise for treating heart attack patients is the endothelial progenitor cell, or EPC. EPCs were discovered by Takayuki Asahara in 1997,[61] and, though they reside in bone marrow, they also circulate throughout the blood. While

58. Lim et al., "Stem Cells Transduced with AKT," 530–42; Li et al., "Bcl-2 Engineered MSCs," 2118–27.

59. Grauss et al., "Forced Myocardin Expression," 1083–93; Song et al., "Modification of Mesenchymal Stem Cells for Cardiac Regeneration," 309–19.

60. Lu et al., "Preconditioning of Stem Cells for Myocardial Infarction," 378–84; Cui et al., "Mesenchymal Stem Cells Preconditioned with Diazoxide," 139–47; Gyöngyösi et al., "Ischaemic Preconditioning on Mesenchymal Stem Cells," 376–84; Xie et al., "Mesenchymal Stem Cells Preconditioned with Hydrogen Sulfide," 29–36.

61. Asahari et al., "Progenitor Endothelial Cells for Angiogenesis," 964–66.

there is a robust debate in the research community as to what distinguishes EPCs from other cells,[62] they can be readily isolated from bone marrow and used in animal and clinical and experiments.[63]

EPCs can restore circulation to damaged animal limbs and damaged animal hearts.[64] Phase 1 clinical trials have demonstrated the safety and feasibility of directly injecting EPCs into the scar tissue of heart attack patients.[65] Phase 2 clinical trials have shown that EPC transplantation into the hearts of patients with congestive heart failure increases circulation in the heart, decreases the symptoms of congestive heart failure, and improves the overall health of the heart.[66]

The drawback of EPCs is that the quality and number of EPCs decreases with age and after a heart attack.[67] However, clinical trials suggest that receiving infusions of EPCs from unrelated people is feasible.[68] Larger studies are in the works,[69] and EPCs definitely have a future in heart treatment either as a stand-alone treatment or in combination with other stem cells.[70]

62. Urbich and Dimmeler, "Endothelial Progenitor Cells," 343–53.

63. Yang et al., "CD34+ Cells."

64. Kocher et al., "Angioblasts," 430–36; Kawamoto et al., "Autologous Endothelial Progenitor Cells," 461–68; Jujo et al., "Endothelial Progenitor Cells in Neovascularization of Infarcted Myocardium," 530–44.

65. Stamm et al., "Myocardial Regeneration," 45–46; Stamm et al., "CD133+ Bone Marrow Cells," 717–25; Klein et al., "Intramyocardial Implantation of CD133+ Stem Cells," E66–69; Ahmadi et al., "Local Autologous Transplantation of CD133+ Enriched Bone Marrow Cells," 153–60.

66. Flores-Ramírez et al., "CD133+ Endothelial Progenitor Cells," 72–78; Losordo et al., "CD34+ Cell Therapy for Refractory Angina," 428–36; Colombo et al., "CD133+ Cell Injection," 239–48.

67. Williamson et al., "Endothelial Progenitor Cells Enter the Aging Arena," 30.

68. Buratovich "Stemedica Ischemic Tolerant Stem Cells Generate Significant Improvement In Ejection Fraction After Heart Attacks," *Beyond the Dish* (blog),April 6, 2012. Online: http://beyondthedish.wordpress.com/2012/04/06/stemedica-ischemic-tolerant-stem-cells-generate-significant-improvement-in-ejection-fraction-after-heart-attacks.

69. Mansour et al., "COMPARE-AMI Trial," 153–59.

70. Bonaros et al., "Combined Transplantation," 1321–28.

Cardiac Stem Cells

In 2001, workers in Michael Rudnicki's laboratory at the University of Ottawa discovered that the heart has its own stem cell.[71] That's right—this organ that no one thought could ever regenerate itself actually regenerates itself constantly. A heart attack simply overtaxes the repair system, and scars form faster than new heart muscle.

Can we help the heart get better at healing itself? This remains one of the most intriguing questions in regenerative medicine. Cardiac stem cells have been heavily studied,[72] and there are several different types of cardiac stem cells.[73] They respond to heart damage, zero in on damaged heart tissue, and divide and replenish dying heart muscle cells.[74] Techniques also exist for isolating and expanding cardiac stem cells not just from mice but from humans as well.[75] Heart biopsies from human patients form little balls of cells called cardiospheres. Transplantation of these human cells into mice with damaged hearts greatly helps them recuperate.[76] Even more fascinating, cardiac stem cells are replenished by special bone marrow cells that move through the bloodstream to the heart.[77]

Cardiac stem cells show a marked ability to repair a damaged heart.[78] In animal experiments, cardiac stem cells were collected from human hearts via biopsies and transplanted into an animal heart after a heart attack. Such experiments in mice,[79] rats,[80] and pigs[81] have shown that cardiac stem cells

71. Hierlihya et al., "Heart Contains a Myocardial Stem Cell Population," 239–43.

72. Oh et al., "Cardiac Progenitor Cells from Adult Myocardium," 12313–18.

73. Leri et al., "Role of Cardiac Stem Cells," 941–61; Buratovich, "Different Kinds of Stem Cells in the Heart," *Beyond the Dish* (blog), March 12, 2012. Online: http://beyondthedish.wordpress.com/2012/03/12/different-kinds-of-stem-cells-in-the-heart.

74. Linke et al., "Stem Cells in the Dog Heart," 8966–71; Beltrami et al., "Adult Cardiac Stem Cells," 763–76.

75. Messina et al., "Cardiac Stem Cells from Human and Murine Heart," 911–21.

76. Smith et al., "Cardiosphere-Derived Cells," 896–908; Davis et al., "Human Cardiospheres," 903–4.

77. Kajstura et al., "Bone Marrow Cells Differentiate in Cardiac Cell Lineages," 127–37; Kucia et al., "Cells Expressing Early Cardiac Markers," 1191–99. This paper is one in a series of publications from this laboratory that proposes that tissue-specific stem cells come from bone marrow.

78. Barile et al., "Endogenous Cardiac Stem Cells," 31–48.

79. Smith et al., "Cardiosphere-Derived Cells," 896–908.

80. Bearzi et al., "Human Cardiac Stem Cells," 14068–73.

81. Johnston et al., "Autologous Cardiosphere-Derived Cells," 1075–83.

become heart muscle that connects with the rest of the heart, shrinks the scar, and improves heart function. Two Phase 1 clinical trials, SCIPIO and CADUCEUS, have produced remarkably encouraging results.[82] In these clinical trials the implanted cardiac stem cells replaced dead heart muscle cells and greatly improved heart function. Other such clinical trials are in progress. When it comes to heart repair, these cells will almost certainly engender tremendous excitement in the years to come.

Umbilical Stem Cells

Umbilical cord and placenta possess several stem cell types that show immense clinical promise. To that end, scientists from American biotechnology companies and Zhengzhou University in China have used MSCs from umbilical cord blood and placenta to treat a patient with congestive heart failure. The patient showed significant improvement after the treatment and exhibited no symptoms of heart failure.[83] While this single trial is not enough to establish the effectiveness of cord blood and placental MSCs as a heart treatment, this is certainly an encouraging result.

Embryonic Stem Cells

Now we come to the $64,000 question: what about embryonic stem cells and fixing the heart? Cultured human or mouse embryonic stem cells can form embryoid bodies that contain groups of cells that spontaneously beat.[84] These beating cells are young heart muscle cells. Skilled scientists can actually manually isolate these beating cells from the other cells. Also, the treat-

82. SCIPIO: Bolli et al., "Cardiac Stem Cells in Patients with Ischaemic Cardiomyopathy (SCIPIO)," 1847–57; Buratovich, "SCIPIO Clinical Trial Shows Remarkable Promise." *Beyond the Dish* (blog), February 2, 2012. Online: https://beyondthedish.wordpress.com/2012/02/02/scipio-clinical-trial-shows-remarkable-promise. CADUCEUS: Makkar et al., "Intracoronary Cardiosphere-Derived Cells for Heart Regeneration after Myocardial Infarction (CADUCEUS)," 895–904; Buratovich, "CADUCEUS Clinical Trial Shows that Cardiosphere-Derived Stem Cells Can Regrow Heart Muscle after a Heart Attack," *Beyond the Dish* (blog), February 15, 2012. Online: http://beyondthedish.wordpress.com/2012/02/15/996.

83. Ichim et al., "Cord Blood Stem Cell Therapy for Dilated Cardiomyopathy," 898–905.

84. Mouse ESCs: Doestschman et al., "Blastocyst-Derived Embryonic Stem Cell Lines," 27–45; Human ESCs: Kehat et al., "Human Embryonic Stem Cells," 407–14.

ment of embryonic stem cell lines with particular chemicals can drive some of the cells to become heart muscle cells.[85] Significant improvements in the techniques for isolating heart muscle cells from the remaining cells have also made it possible to make pure heart muscle cultures from embryonic stem cells.[86]

Glossary

Embryoid bodies – clumps of cells derived from embryonic stem cells. Extensive and uncontrolled differentiation occurs in these clumps, and these clumps contain multiple cell types. Clinical use of such cells required purification of the desired cells from the clumps, or directing the differentiation with various combinations of particular chemicals that drive the cells to form one cell type over another.

Several labs have transplanted heart muscle cells made from embryonic stem cells in animal studies. In these experiments, laboratory animals are given heart attacks, followed by transplantation of heart muscle cells made from embryonic stem cells into the heart. However, oftentimes the embryonic stem cell–derived heart muscle cells come from a different animal species than the heart-sick laboratory animals into which they are transplanted.[87] While this seems a rather strange way to do science, it results from the fact that some robust embryonic stem cell lines work well for heart-based experiments and some types of laboratory animals serve as fine model systems for heart attacks and other cardiac infirmities. Therefore, though not completely ideal, this experimental strategy works quite well.

85. Yao et al., "Differentiation of Human Embryonic Stem Cells," 6907–12; Laflamme et al., "Cardiomyocytes Derived from Human Embryonic Stem Cells," 1015–24; Wang et al., "Cardiac Induction of Embryonic Stem Cells," 192–97; Zhang et al., "Direct Differentiation of Atrial and Ventricular Myocytes," 579–87.

86. Van Hoof et al., "Antibody-Based Selection," 1610–18.

87. For all xenotransplantation (transplantation from one species into another) experiments, laboratory animals are either treated with immunosuppressive drugs or possess genetic conditions that prevent them from rejecting foreign transplants. See Zhang et al., "Usages of the 'Nude' Mouse," 1156–67.

What have such heart muscle transplantation experiments shown? When laboratory rats that have suffered a heart attack receive heart muscles derived from mouse embryonic stem cells, these implants preserve the structure of the heart, shrink the scar, and prevent cell death. Also, the electrical profile of the heart as measured by electrocardiograms (ECG or EKG) becomes more normal.[88] Experiments similar to the one described above have transplanted heart muscle cells made from human embryonic stem cells into laboratory rats[89] and monkeys,[90] and mouse embryonic stem-derived heart muscle into sheep,[91] all with quite positive results.

Despite these successes some concerns have surfaced. One is that most of these studies have examined the effects only after four weeks. Studies in rats with heart muscle made from mouse embryonic stem cells have demonstrated long-term improvements in heart function after transplantation.[92] However, a longer-term study that used heart muscle made from human embryonic stem cells in rats showed initial improvements in heart function four weeks after transplantation, but failed to show sustained improvements in the heart at three months.[93] Therefore, the benefits of implanted embryonic stem cell-derived heart muscle might be short-lived.

A further concern is that tumors form when incompletely differentiated embryonic stem cells are transplanted.[94] In the heart, such tumors are quite nasty and can prove rather dangerous to the patient. Several studies have documented the formation of tumors in the heart if incompletely differentiated embryonic stem cells are transplanted.[95] This is potentially

88. Hodgson et al., "Stable Benefit of Embryonic Stem Cell Therapy," H471–79; Min et al., "Transplantation of Embryonic Stem Cells," 288–96; Qiao et al., "Long-Term Improvement," 33–41.

89. Laflamme et al., "Formation of Human Myocardium," 663–71; Caspi et al., "Transplantation of Human Embryonic Stem Cell–Derived Cardiomyocytes," 1884–93.

90. Bin et al., "Population of Multipotent Cardiovascular Progenitors," 1125–39.

91 Menard et al., "Cardiac-Committed Mouse Embryonic Stem Cells," 1005–12.

92 Hodgson et al., "Stable Benefit of Embryonic Stem Cell Therapy," H471–79; Qiao et al., "Long-Term Improvement," 33–41.

93. van Laake et al., "Cardiac Repair in Rodents," 1008–10.

94. Some researchers suspected that after a heart attack the chemicals produced by the damaged heart could direct implanted embryonic stem cells to differentiate into heart cells. Experiments have shown that this is not the case. See Chen et al., "Infarcted Cardiac Microenvironment," 77–83; He et al., "Fate of Undifferentiated Mouse Embryonic Stem Cells," 188–201.

95. Caspi et al., "Transplantation of Human Embryonic Stem Cell–Derived Cardiomyocytes," 1884–93; Nussbaum et al., "Teratoma Formation and Immune Response," 1345–57.

worrisome; however, improved enrichment techniques allow researchers to isolate pure cultures of heart muscle cells from embryonic stem cell cultures, thus largely erasing the problem of tumor formation.[96]

A third concern is the tendency for embryonic stem cells to cause hearts to beat abnormally when transplanted into adult hearts. Normally the heartbeat begins in the upper right chamber of the heart. The upper chambers of the heart are called atria, and each heartbeat starts with the generation of an electrical impulse by a small island of tissue in the right atrium called the pacemaker or sinus node. This electrical impulse spreads to the left atrium, which causes the two upper chambers to contract simultaneously and propel blood from the upper chambers of the heart to the lower chambers. The beat impulse is relayed to the lower chambers (ventricles) by a conductive system that acts as a sort of "extension cord" from the upper part of the heart to the lower part. Once the signal is relayed by this extension cord, the beat impulse spreads throughout the heart muscle, and the lower chambers of the heart contract. Contractions of the lower chambers of the heart propel blood into the large blood vessels that take blood to the lungs (right side of the heart) or to the rest of the body (left side).

If the conduction of the beat impulse goes awry, then irregular heartbeats can ensue. If the pacemaker is damaged, or if the beat impulse is partially blocked on the way down to the lower chambers of the heart, then the heart will usually beat too slowly (less than sixty beats per minute), a condition called bradycardia. Too slow a heartbeat can increase one's risk of having a heart attack or stroke. Too rapid a heart rate (over one hundred beats per minute, called tachycardia) occurs if tissue other than the pacemaker in the right atrium starts to initiate heartbeats. A heart that beats too rapidly requires more oxygen, and if the body cannot deliver sufficient oxygen to the faster-working heart, the heart muscle will start to die.

A kind of chaotic beat can occur if another portion of the heart asserts itself as the new manager of the heartbeat. In this case, the heart muscle receives conflicting instructions as to when to contract.[97] This is the equivalent of an orchestra having more than one conductor; the musicians

96. Lin et al., "Tumorigenesis in the Infarcted Rat Heart," 1179–85.

97. A part of the heart that generates a beat impulse without waiting for the sinus node to do it first is called an ectopic focus. Ectopic foci can cause a single premature heartbeat every so often, but they can also cause sustained abnormal heart rhythms if they fire more often than the cells of the sinus node. Ectopic foci in the atria are not nearly as dangerous as those in the ventricles.

have no idea whom they should follow, and the result is chaos rather than harmony. In addition, if the heart contains dead tissue and, consequently, the movement of the beat impulse is slower in one region than in the rest of the heart, a delayed beat impulse can be interpreted by the resident heart muscle as a new beat.[98]

This type of chaos is dangerous to the heart. If the pacemaker is not the only tissue setting the beat impulse, the heart muscle does not follow one signal, but multiple signals. The top of the heart must contract before the bottom of the heart. If this orderly contraction does not occur, the heart can begin to throw off premature beats or beat irregularly. If the chaos in the heart becomes severe, then the heart muscle will quiver instead of beat. A quivering heart does not effectively pump blood throughout the body.[99]

Therefore one of the main concerns with embryonic stem cell–derived heart treatments is that heart muscle made from embryonic stem cells typically acts like very young heart muscle.[100] Several physiological studies of heart muscle cells made from embryonic stem cells have confirmed that these cells are more like immature heart muscle cells.[101] The implantation of young heart muscle into the heart of an older person causes a mismatch between the heart muscle cells. Younger hearts tend to beat very fast, while older hearts beat more slowly. The result is an irregular heartbeat, and implanted heart muscle cells derived from embryonic stem cells have the

98. This phenomenon is called reentry. Reentry occurs if an electrical impulse recurrently travels in a tight circle within the heart rather than stopping after moving from one end of the heart to the other. Each heart cell can transmit impulses in every direction, but can only do so once within a short period of time. Normally an impulse spreads through the heart so fast that heart cells only respond once. However, if conduction of the beat impulse is abnormally slow in some areas, part of the impulse arrives late and is treated as a new impulse. Depending on the timing, this can produce a sustained abnormal circuit rhythm. Reentry circuits are responsible for pathologies such as atrial flutter, most cases of paroxysmal supraventricular tachycardia, and dangerous ventricular tachycardia.

99. When an entire chamber of the heart is involved in a multiple reentry circuit it quivers with chaotic electrical impulses and is in fibrillation. Atrial fibrillation affects the upper chambers of the heart and is not necessarily a medical emergency. Ventricular fibrillation is always a medical emergency.

100. He et al., "Multiple Types of Cardiac Myocytes," 31–39; Cao et al., "Human Embryonic Stem Cell-Derived Cardiomyocytes."

101. Cardiomyocytes derived from human ESCs lack key intracellular Ca2+ handling proteins (calsequestrin and phospholamban). See Binah et al., "Functional and Developmental Properties," S192–96. Additionally, cardiomyocytes made from ESCs "expressed cardiomyocyte genes at levels similar to those found in 20-week fetal heart cells." See Cao et al., "Transcriptional and Functional Profiling."

capability to generate heart arrhythmias that are sometimes severe.[102] This problem could potentially make the cure worse than the disease.

A final concern with cells derived from embryonic stem cells is that they possess the surface proteins encoded by the genes found in the original embryonic stem cells. If these surface proteins are different from those found in the patient, the patient's immune system will identify the transplanted tissue as foreign and will attack and destroy it. Several animal experiments have shown that despite early hopes, tissues made from embryonic stem cells are recognized as foreign by the immune system and destroyed.[103]

While a great deal of hope is placed in embryonic stem cells, substantial hurdles still exist before they can be used in clinical trials.

Induced Pluripotent Stem Cells

Several labs have successfully converted induced pluripotent stem cells (iPSCs) into heart muscle cells.[104] Heart muscle cells made from iPSCs can also repair the hearts of lab animals that have had a heart attack, without signs of tumor formation.[105] In terms of their electrical profiles, iPSC-derived heart muscle cells appear to be more mature than those derived from embryonic stem cells.[106] Nevertheless, the safety of iPSC lines differs significantly from line to line,[107] and iPSCs must undergo much more testing before they prove themselves safe for human clinical trials. However, iPSCs can model heart diseases in a very accurate way and patient-specific modeling of heart diseases with iPSCs has already saved at least one person's life.[108]

102. Zhang et al., "Arrhythmic Potential," 1294–99; Liao et al., "Proarrhythmic Risk of Embryonic Stem Cell–Derived Cardiomyocyte," 1853–59.

103. Swijnenburg et al., "Embryonic Stem Cell Immunogenicity Increases," I166–72; Kofidis et al., "Embryonic Stem Cells Trigger Host Immune Response," 461–66.

104. Burridge et al., "Highly Efficient Cardiac Differentiation"; Rajala et al., "Cardiac Differentiation of Pluripotent Stem Cells," 1–12.

105. Nelson et al., "Repair of Acute Myocardial Infarction," 408–16; Singla et al., "iPS Cells Repair and Regenerate Infarcted Myocardium," 1573–81.

106. Itzhaki et al., "Calcium Handling"; Poon et al., "Electrophysiological Perspective," 1495–1504.

107. Miura et al., "Safety of Induced Pluripotent Stem Cell Lines," 743–45.

108. See Buratovich, *Beyond the Dish* (blog), January 21, 2013, http://beyondthedish.wordpress.com/2013/01/21/induced-pluripotent-stem-cells-used-to-define-proper-treatment-for-heart-patient.

Unlike embryonic stem cells, adult stem cells are actually curing weak hearts, and start-up companies have been offering adult stem cell treatments outside the United States for several years. The problem with present adult stem cell treatments is that they show inconsistent results, helping some people a great deal but helping others not at all.

There you have it. Can stem cells help an ailing heart? The answer is a clear-cut yes!

Praying for you,

Michael Buratovich

Letter #30

The Bottom Line

Dear Dr. Buratovich,

Our daughter Jennifer came home for spring break and she had all these notes and pictures about stem cells. She said that she had learned a lot about the subject in your seminar class. She started talking at us at about a hundred miles a minute, and my husband's eyes and mine glazed over after about two minutes. We couldn't keep up with her. She has majored in science and is definitely past our scientific levels of education.

Mike, I'm a paralegal and my husband is a businessman. We are not stupid, but we do not have the science education that Jennifer has or, especially, that you have. I have read the newspapers, and I do not believe half of what I read these days, but I am still desperately interested in this question.

Can you give it to me straight and put the cookies on the lowest shelf for me? Skip all the science lingo and jargon. What are the bare facts on this issue? Can you tell me what the bottom lines are when it comes to the embryonic stem cell debate?

Janet S. H.

Dear Janet,

I do not usually get letters from parents, and when I do, it is usually about their student's grade. It was a very pleasant surprise to see a letter from a parent about the subject of my class. This tickled me pink.

So you want the bottom line? I can do that, but it might be a bit of a whirlwind. Strap on your helmet. Here we go.

In the first place, human embryos are young human lives and are in fact human persons. Embryos are not yet conscious; they have no feelings to hurt or opinions to state. Nevertheless, if we were to have a scrapbook of our childhood, it could start (particularly if we were conceived through in vitro fertilization) with a picture of the embryo that was transferred into the mother's womb. There is continuity from the embryo to the fetus to the newborn to the toddler and so on. Arguments that assert that the embryo is not a human person because the embryo is not aware or rational should equally apply to newborn babies, but we protect those. Such protections should apply to the embryo as well.

Second, making embryonic stem cells kills human embryos. The papers will say things like "it kills blastocysts," but a blastocyst is a human embryo at the blastocyst stage. Don't fall for this double-talk. The classical way of making embryonic stem cells (ESCs) kills young human persons. Also remember that "Snowflake" children were all adopted as embryos. Excess embryos made at fertility clinics by in vitro fertilization are either implanted into the mother's womb or frozen for safekeeping. These stored embryos are eventually discarded, donated to research, or given up for adoption. Children whose births result from embryo adoption are called Snowflake children, and from 2004 to 2006, there were 988 babies born who were originally adopted as embryos. However, since the founding of the Snowflakes Frozen Embryo Adoption Program in 1997, the number of children born from adopted embryos might now exceed three thousand.[1] Consider that one embryo is adopted and produces a bouncing baby, while another equally fine embryo goes to a laboratory and is destroyed. Why should one suffer death and dismemberment but the other continue to live and develop? Clearly embryos are immature human persons who deserve protection under the law.

Third, to date, there are no ESC-based treatments, despite the buckets and buckets of money that have been thrown at this research, and all the hype. When ESC lines were first published, newspapers predicted

1. Lester, "Embryo Adoption Becoming the Rage."

ESC-based cures within ten years.[2] We are more than fourteen years from the isolation of ESCs and we have two clinical trials to show for it. The first of these clinical trials was conducted by Geron Corporation, which treated at least four acute spinal cord injured patients with embryonic stem cell–derived nerve-insulating cells (oligodendrocyte precursor cells). While these implanted cells caused no severe adverse side effects, their patients showed so little improvement that Geron discontinued funding for their embryonic stem cell division.[3] In another Phase 1/2 clinical trial, Advanced Cell Technology received FDA approval to treat two eye patients with retinal pigment epithelial cells derived from ESCs. One patient showed definite structural improvement in the injected eye and measurable increases in optical acuity. However, the other patient showed marginal improvements in the eye that had been treated and in the untreated eye as well.[4]

These results are hardly overwhelming. We can certainly be glad for the patient who showed improvement, but of these two clinical trials, one ended prematurely and with nothing to show for it and the other is extremely small and largely inconclusive. These results show that ESCs have lived up to neither their promise nor their hype. Instead, they show that making safe cells that can treat particular maladies from embryonic stem cells is a very tough problem. Furthermore, more and more people every day have been successfully treated with adult and umbilical cord stem cells, and the available procedures and treatments are growing. This is a simple, undeniable fact. Now, proponents of ESCs will say that the adult stem cell field had a massive head start since bone marrow transplants were developed in the 1950s and 1960s. That is only half true, since bone marrow transplants have been performed only since the 1980s, and the first human adult stem was isolated in 1992.[5] This shows that adult stem cells present fewer obstacles for developing treatments, and this is another reason why

2. Perlman, "Cures within Next 10 Years."

3. Smith, "Geron Abandons Embryonic Stem Cell Research"; Buratovich, "More on Geron Leaving the Embryonic Stem Cell Business," *Beyond the Dish* (blog), 11/17/2011. Online: http://beyondthedish.wordpress.com/2011/11/17/more-on-geron-leaving-the-embryonic-stem-cell-business.

4. The patient who experienced improvement had Stargardt disease, and the other patient who showed limited improvements in both eyes had age-related macular degeneration. See Schwartz, Hubschman, and Heilwell, "Embryonic Stem Cell Trials for Macular Degeneration," 713–20; Buratovich, "Stem Cells Make the Blind (Lab Animals) See," *Beyond the Dish* (blog). Online: http://beyondthedish.wordpress.com/2012/02/17/bone-marrow-stem-cells-make-the-blind-lab-animals-see.

5. Baum et al., "Isolation of Human Hematopoietic Stem-Cell," 2804–8.

adult stem cells have taken the regenerative treatment field by storm while ESCs are still lagging at the starting gate.

Fourth, all the objections regarding the personhood of the embryo have to do with its development. The high death rate of embryos under natural conditions, the lack of adhesion or interaction between cells of the early embryo, twinning, embryo recombination, molar pregnancies, and so on are all characteristics or consequences of either normal or abnormal development. These arguments fault human embryos for going through their usual developmental processes, or they question the personhood of all embryos because in some of them the developmental process can go awry. None of these are good reasons to question the personhood of embryos.

Fifth, if people tell you that the Christian church allowed abortion before quickening for most of its history, they are simply misinformed. The earliest pronouncements on abortion by Christians make it abundantly clear that Christians held that abortion was wrong at any stage of development and this was the original view of the Christian church. The formed/unformed distinction came later, and the two views actually existed side by side for most of the history of the medieval Catholic Church. Also, in order to fulfill what they believed was a God-given directive, the early Christians sought to rescue, protect, and nurture those who were weak and defenseless against the Greco-Roman culture of death. The unborn fell right into this category regardless of their age, and defending the human life at this earliest part of life when it is most vulnerable would have been perfectly consistent with the behavior and practice of the early church.

Sixth, cloned embryos are human persons who deserve protection under the law. Making cloned embryos in order to destroy them for their cells is simply immoral. In addition, making cloned embryos requires that you persuade women to donate their eggs for the procedure, and this has harmed women and provided ample opportunities for exploitation of poor women.

Seventh, cloning for reproduction is just plain wrong. It puts the cloned child, who is regarded as a manufactured product rather than a human person, in a lose-lose situation. She is made to fulfill a specific expectation or to replace a specific person, but because cloning does not produce an exact carbon copy of the parent organism, she almost certainly will not be like that person she is supposed to replace. Therefore the clone is valued for what she can do rather than who she is. No one should have to live like that, and such a situation is inherently cruel.

Eighth, of the surplus embryos made by in vitro fertilization, the vast majority of them in the United States and Europe belong to couples who wish to use them to build their families. Those families who have finished having children and have extra frozen embryos feel a great deal of anxiety over what to do with them because they understand the simple truth that reproductive physicians refuse to recognize: human embryos are very young human persons. Medical embryology textbooks recognize this simple fact. Consider the medical embryology textbook by O'Rahilly and Muller, which states, "Although life is a continuous process, fertilization is a critical landmark because, under ordinary circumstances, a new, genetically distinct human organism is thereby formed. . . . The combination of 23 chromosomes present in each pronucleus results in 46 chromosomes in the *zygote* [after the egg is fertilized it is called a zygote]. Thus the diploid number [i.e., two copies of each chromosome] is restored and the embryonic genome is formed. The embryo now exists as a genetic unity."[6] Another embryology textbook by Moore and Persuad states, "This cell results from the union of an oocyte (egg) and a sperm during fertilization. A zygote or embryo is the beginning of a new human being."[7] To get around this simple fact, reproductive physicians coined a pseudoscientific term for the early embryo: the "pre-embryo." Not only is this term completely meaningless from an embryological perspective,[8] but embryonic stem cell advocate Lee Silver admits that "pre-embryo" was coined solely for political reasons:

> I'll let you in on a secret. The term *pre-embryo* has been embraced wholeheartedly by IVF practitioners for reasons that are political, not scientific. The new term is used to provide the illusion that there is something profoundly different between what we nonmedical biologists still call a six-day-old embryo and what we and everyone else call a sixteen-day-old embryo.[9]

The large quantity of frozen embryos in the United States is a testimony to the absence of meaningful regulation of the fertility industry and the deplorable lack of wisdom our nation has exercised in the use and promulgation of artificial reproductive technologies.

Ninth, you do not need to destroy embryos to make embryonic stem cells. While induced pluripotent stem cells (iPSCs) are not exactly like ESCs

6. O'Rahilly and Muller, *Human Embryology and Teratology*, 8, 29.

7. Moore and Persuad, *Developing Human*, 2.

8. O'Rahilly and Muller, *Human Embryology and Teratology*, 12. They call this term "ill-defined and inaccurate."

9. Silver, *Remaking Eden*, 39.

in every respect, the simple truth is that iPSCs are similar enough to ESCs in so many essential ways that they can certainly fulfill the roles of ESCs.[10] We do not need to destroy embryos to make these cells, and we should simply use the available lines and cease all embryo destruction. Additionally, the September 1999 report issued by President Clinton's National Bioethics Advisory Commission (NBAC) concluded that "the derivation of stem cells from embryos remaining following infertility treatments is justifiable only if not less morally problematic alternatives are available for advancing the research. . . . The claim that there are alternatives to using stem cells derived from embryos is not, at the present time, supported scientifically. We recognize, however, that this is a matter that must be revisited continually as science advances."[11] Despite all the advances made with iPSCs and adult stem cell treatment, ESC research still remains.

Tenth, if a politician makes a promise about stem cells, don't believe him. In 2004 Democratic vice presidential candidate John Edwards even went so far as to promise that if he and John Kerry were elected, "people like Christopher Reeves are going to walk, get up out of that wheelchair and walk again."[12] This wasn't true in 2004, and it isn't true today. This was a foolish thing to say, but politicians say things to get people to like them. Once people stop liking them they lose their jobs, and the truth or proper nuance of a subject often gets lost in the world of talking points and sound bites.

Finally, keep your wits about you. If you engage people on this subject and you disagree with them, be prepared to be called every name in the book. You are not going to convince some people. That's fine—just move on and engage someone who might actually listen.

Those are the main and plain things on this issue. Now get out there and argue your point of view in op-ed pieces, interviews, and in the workplace. People change their minds when others personally articulate their views in a kind but informed manner. Be that ambassador for the youngest and weakest of us who cannot yet speak up for themselves!

Cheers and God bless!

Michael Buratovich

10. Buratovich, "Protein Production," *Beyond the Dish* (blog), September 14, 2011. Online: https://beyondthedish.wordpress.com/2011/09/14/induced-pluripotent-stem-cells-are-almost-exactly-like-embryonic-stem-cells-when-it-comes-to-protein-production.

11. National Bioethics Advisory Commission, *Ethical Issues in Human Stem Cell Research*, vol 1.

12. Krauthammer, "Edwards Outrage."

Epilogue

An old Indian parable tells of a wealthy and wise jewel merchant who was met on a journey by an old acquaintance who insisted that they travel together. Because the jewel merchant was uncertain of the motives of his newfound traveling companion, he resolved never to leave his precious stones alone with him. Therefore, when it was his turn to wash before sleeping, the jewel merchant hid his bag of precious stones under the pillow of his traveling companion. While he was washing, the jewel merchant's bed and luggage were ransacked by his greedy traveling companion. However, every night his search was in vain. At the end of their journey together, the jewel merchant revealed to his old acquaintance that the treasure he deeply coveted had lain underneath his own pillow. Had he looked in his own bed, he would have found what he wanted. "The wealth was nearer to you than you realized," the wise merchant told his astonished, avaricious acquaintance.[1]

Many experiments have definitively shown that regenerative therapy has a distinct ability to cure sick people. A wide variety of cells can help heal a damaged heart, reconstruct someone's damaged or cancerous bone marrow, and even possibly cure diabetes. Like the traveler who looked everywhere, but in all the wrong places, we are looking in the wrong places for cures. People are walking among us whose failing hearts have been renewed by adult and umbilical cord stem cell transplants. They are curing people right now. While we should continue research on regenerative medicine with the available cell lines, we should also remember to invest the majority of our resources into what works.

1. Zacharias and Fournier, *The Merchant and the Thief*, 1–32

Bibliography

Abdel-Latif, Ahmed, Roberto Bolli, Imad M. Tleyjeh, et al. "Adult Bone Marrow-Derived Cells for Cardiac Repair: A Systematic Review and Meta-Analysis." *Archives of Internal Medicine* 167, no. 10 (2007): 989–97.

Adler, Eric D., Vincent C. Chen, Anne Bystrup, et al. "The Cardiomyocyte Lineage is Critical Optimization of Stem Cell Therapy in a Mouse Model of Myocardial Infarction." *The FASEB Journal* 24 (2010): 1073–81.

Ahmad, Ruhel, Wanja Wolber, Sigrid Eckardt, et al. "Functional Neuronal Cells Generated by Human Parthenogenetic Stem Cells." *PLoS One* 7 (2012): e4125.

Ahmadi, Hossein, Hossein Baharvand, Saeed Kazemi Ashtiani, et al. "Safety Analysis and Improved Cardiac Function Following Local Autologous Transplantation of CD133+ Enriched Bone Marrow Cells After Myocardial Infarction." *Current Neurovascular Research* 4, no. 3 (2007): 153–60.

Ahn, Tae-Beom, J. William Langston, Venkat Raghav Aachi, and Dennis W. Dickson. "Relationship of Neighboring Tissue and Gliosis to α-Synuclein Pathology in a Fetal Transplant for Parkinson's Disease." *American Journal of Neurodegenerative Disease* 1 (2012): 49–59.

Ailles, Lautie E., and Luigi Naldini. "HIV-1-Derived Lentivirus Vectors." *Current Topics in Microbiology and Immunology* 261 (2002): 31–52.

Alipio, Zaida, Wenbin Liao, Elizabeth J. Roemer, et al. "Reversal of Hyperglycemia in Diabetic Mouse Models Using Induced-Pluripotent Stem (iPS)-Derived Pancreatic β-Like Cells." *Proceedings of the National Academy of Sciences USA* 107, no. 30 (2010): 13426–31.

Allen, Nicholas D., Sheila C. Barton, Kathy Hilton, Mike L. Norris, and M. Azim Surani. "A Functional Analysis of Imprinting in Parthenogenetic Embryonic Stem Cells." *Development* 120 (1994): 1473–82.

Althuis, Michelle D., Kamran S. Moghissi, Carolyn L. Westhoff, et al. "Uterine Cancer After Use of Clomiphene Citrate to Induce Ovulation." *American Journal of Epidemiology* 161, no. 7 (2005): 607–15.

Ambrose. *Hexameron, Paradise and Cain and Abel.* Translated by John J. Savage. Vol. 42 of *The Fathers of the Church*, edited by Hermigild Dressler, Robert P. Russell, Thomas P. Halton, William R. Tongue, and Josephine Brennan, 207–8. Washington D. C.: Catholic University of America Press, 1961.

Amit, Michal, Melissa K. Carpenter, Margaret S. Inokuma, et al. "Clonally Derived Human Embryonic Stem Cell Lines Maintain Pluripotency and Proliferative Potential for Prolonged Periods of Culture." *Developmental Biology* 227, no. 2 (2000): 271–8.

Amoh, Yasuyuki, Lingna Li, Kensei Katsuoka, and Robert M. Hoffman. "Multipotent Hair Follicle Stem Cells Promote Repair of Spinal Cord Injury and Recovery of Walking Function." *Cell Cycle* 7, no. 12 (2008): 1865–9.

Amoh, Yasuyuki, Maho Kanoh, Shiro Niiyama, et al. "Human Hair Follicle Pluripotent Stem (hfPS) Cells Promote Regeneration of Peripheral-Nerve Injury: An

Advantageous Alternative to ES and iPS Cells." *Journal of Cellular Biochemistry* 107, no. 5 (2009):1016–20.

Amps, K., P. W. Andrews, G. Anyfantis, et al. "Screening Ethnically Diverse Human Embryonic Stem Cells Identifies a Chromosome 20 Minimal Amplicon Conferring Growth Advantage." *Nature Biotechnology* 29, no. 12 (2011): 32–44.

Amrolia, Persis, Hubert B. Gaspar, Amel Hassan, et al. "Nonmyeloablative Stem Cell Transplantation for Congenital Immunodeficiencies." *Blood* 96, no. 4 (2000): 1239–46.

Anderson, David F., P. Prabhasawat, E. Alfonso, and Scheffer C. G. Tsenga. "Amniotic Membrane Transplantation after the Primary Surgical Management of Band Keratopathy." *Cornea* 20, no. 4 (2001): 354–61.

Anderson, David F., Pierre Ellies, Renato T. F. Pires, and Scheffer C. G. Tsenga. "Amniotic Membrane Transplantation for Partial Limbal Stem Cell Deficiency." *British Journal of Ophthalmology* 85 (2001): 567–75.

Antczak, Michael, and Jonathan Van Blerkom. "Oocyte Influences on Early Development: The Regulatory Proteins Leptin and STAT3 are Polarized in Mouse and Human Oocytes and Differentially Distributed within the Cells of the Preimplantation Stage Embryo." *Molecular Human Reproduction* 3 (1997): 1067–86.

Antczak, Michael, and Jonathan Van Blerkom. "Temporal and Spatial Aspects of Fragmentation in Early Human Embryos: Possible Effects on Developmental Competence and Association with the Differential Elimination of Regulatory Proteins from Polarized Domains." *Human Reproduction* 14 (1999): 429–47.

Aoi, Takashi, Kojiro Yae, Masato Nakagawa, et al. "Generation of Pluripotent Stem Cells from Adult Mouse Liver and Stomach Cells." *Science* 321 (2008): 699–702.

Aquinas, Thomas. *Commentary on the Sentences of Peter Lombard: English and Latin*. 8 vols. Bellingham WA: Logos Bible Software, 2013.

Aquinas, Thomas. *Summa Contra Gentiles*. Translated by Vernon J. Bouke. Vol. 3, part 2 of *Summa Contra Gentiles*, edited by A. C. Pegis, J. F. Anderson, V. J. Bourke, and C. J. O'Neil, 142–7. Notre Dame, IN: University of Notre Dame Press, 1975.

Aquinas, Thomas. *Summa Theologica*. Claremont, CA: Coyote Canyon Press, 2010. Kindle edition.

Arima, Takahiro, Takao Matsuda, Nobuo Takagi, and Norio Wake. "Association of IGF2 and H19 Imprinting with Choriocarcinoma Development." *Cancer Genetics and Cytogenetics* 90 (1997): 39–47.

Aristotle. *The Basic Works of Aristotle*. Edited by Richard McKeon. New York: Random House, 1941.

Arnhold, Stefan, S. Glüer, K. Hartmann, et al. "Amniotic-Fluid Stem Cells: Growth Dynamics and Differentiation Potential after a CD-117-Based Selection Procedure." *Stem Cells International* 2011 (2011): 715341. doi:10.4061/2011/715341.

Arnold, Daniel R., Vilceu Bordignon, Réjean Lefebvre, Bruce D. Murphy, and Lawrence C. Smith. "Somatic Cell Nuclear Transfer Alters Peri-Implantation Trophoblast Differentiation in Bovine Embryos." *Reproduction* 132 (2006): 279–90.

Asahari, Takayuki, Toyoaki Murohara, Alison Sullivan, et al. "Isolation of Putative Progenitor Endothelial Cells for Angiogenesis." *Science* 274, no. 5302 (1997): 964–6.

Asanoma, Kazuo. "*NECC1*, A Candidate Choriocarcinoma Suppressor Gene that Encodes a Homeodomain Consensus Motif." *Genomics* 81 (2003): 15–25.

Ashley, Benedict, and Albert Moraczewski. "Is the Biological Subject of Human Rights Present from Conception?" In *The Fetal Issue: Medical and Ethical Aspects*, edited by Peter Cataldo and Albert Moraczewski, 33–60. Braintree, MA: Pope John Center, 1994.

Ashton, Brian A., T. D. Allen, C. R. Howlett, C. C. Eaglesom, A. Hattori, and M. Owen. "Formation of Bone and Cartilage by Marrow Stromal Cells in Diffusion Chambers In Vivo." *Clinical Orthopaedics and Related Research* 151 (1980): 294–307.

Assmus, Birgit, Andreas Rolf, Sandra Erbs, et al. "Clinical Outcome 2 Years after Introcoronary Administration of Bone Marrow-Derived Progenitor Cells in Acute Myocardial Infarction." *Circulation Heat Failure* 3 (2010): 89–96.

Aston, Kenneth I., C. Matthew Peterson, and Douglas T. Carrell. "Monozygotic Twinning Associated with Assisted Reproductive Technologies: A Review." *Reproduction* 136 (2008): 377–86.

Atala, Anthony, Stuart B. Bauer, Shay Soker, James J. Yoo, and Alan B. Retik. "Tissue-Engineered Autologous Bladders for Patients Needing Cystoplasty." *Lancet* 367 (2006): 1241–6.

Atala, Anthony. "Tissue Engineering of Human Bladder." British Medicine Bulletin 97, *no.* 1 *(*2011*):* 81–194.

Austriaco, Nicanor Pier Giorgio. "Altered Nuclear Transfer: A Critique of a Critique." *Communio* 32 (2005): 172–6.

Avellana-Adalid, Virginia, C. Bachelin, F. Lachapelle, et al. "In Vitro and In Vivo Behaviour of NDF-Expanded Monkey Schwann Cells." *European Journal of Neuroscience* 10, no. 1 (1998): 291–300.

Azmanov, Dimitar N., Tania V. Milachich, Boriana M. Zaharieva, et al. "Profile of Chromosomal Aberrations in Different Gestational Age Spontaneous Abortions Detected by Comparative Genomic Hybridization." *European Journal of Obstetrics and Gynecology and Reproductive Biology* 131 (2007): 127–31.

Baguisi, Alexander, Esmail Behboodi, David T. Melican, et al. "Production of Goats by Somatic Cell Nuclear Transfer." *Nature Biotechnology* 17 (1999): 456–61.

Bailey, Mark M., Limin Wang, Claudia J. Bode, Kathy E. Mitchell, and Michael S. Detamore. "A Comparison of Human Umbilical Cord Matrix Stem Cells and Temporomandibular Joint Condylar Chondrocytes for Tissue Engineering Temporomandibular Joint Condylar Cartilage." *Tissue Engineering* 13, no. 8 (2007): 2003–10.

Bailey, Ronald. "Are Stem Cells Babies?" *Reason.com*. July 11, 2001. http://reason.com/archives/2001/07/11/are-stem-cells-babies/print.

Bailo, Marco, Maddalena1 Soncini, Elsa Vertua, et al. "Engraftment Potential of Human Amnion and Chorion Cells Derived from Term Placenta." *Transplantation* 78 (2004): 1439–48.

Balaban, Basak, Bulent Urman, Aycan Sertac, Cengiz Alatas, Senai Aksoy, and Ramazan Mercan. "Blastocyst Quality Affects the Success of Blastocyst-Stage Embryo Transfer." *Fertility and Sterility* 74, no. 2 (2000): 282–7.

Balaban, Basak, Bulent Urman, Cengiz Alatas, Ramazan Mercan, Senai Aksoy, and Aycan Isiklar. "Poor-Quality Cleavage-Stage Embryos Results in Higher Implantation Rates." *Fertility and Sterility* 75, no. 3 (2001): 514–8.

Balaban, Basak, Kayhan Yakin, and Bulent Urman. "Randomized Comparison of Two Different Blastocyst Grading Systems." *Fertility and Sterility* 85, no. 3 (2006): 559–63.

Ballen, Karen K., Juliet N. Barker, Susan K. Stewart, Michael F. Greene, and Thomas A. Lane. "Collection and Preservation of Cord Blood for Personal Use." *Biology of Blood and Marrow Transplantation* 14, no. 3 (2008): 356–63.

Balsam, Leora B., Amy J. Wagers, Julie L. Christensen, Theo Kofidis, Irving L. Weissman, and Robert C. Robbins. "Haematopoietic Stem Cells Adopt Mature Haematopoietic Fates in Ischemic Myocardium." *Nature* 428 (2004): 668–73.

Ban, De-Xiang, Guang-Zhi Ning, Shi-Qing Feng, et al. "Combination of Activated Schwann Cells with Bone Mesenchymal Stem Cells: The Best Cell Strategy for Repair after Spinal Cord Injury in Rats." *Regenerative Medicine* 6, no. 6 (2011): 707–20.

Bankowski, Brandon J., Anne D. Lyerly, Ruth R. Faden, and Edward E. Wallach. "The Social Implications of Embryo Cryopreservation." *Fertility and Sterility* 84, no. 4 (2005): 823–32.

Barile, Lucio, Elisa Messina, Alessandro Giacomello, and Eduardo Marbán. "Endogenous Cardiac Stem Cells." *Progress in Cardiovascular Diseases* 50, no. 1 (2007): 31–48.

Barnabé-Heider, Fanie and Jonas Frisén. "Stem Cells for Spinal Cord Repair." *Cell Stem Cell* 3 (2008): 16–24.

Barnett, Susan C., and Lynda Chang. "Olfactory Ensheathing Cells and CNS Repair: Going Solo or in Need of a Friend?" *Trends in Neuroscience* 27 (2004): 54–60.

Barratt, Christopher L. R., Justin C. St. John, and Masoud Afnan. "Clinical Challenges in Providing Embryos for Stem-Cell Initiatives." *Lancet* 364, no. 9429 (2004): 115–8.

Bartholomew, Amelia, Cord Sturgeon, Mandy Siatskas, et al. "Mesenchymal Stem Cells Suppress Lymphocyte Proliferation In Vitro and Prolong Skin Graft Survival In Vivo." *Experimental Hematology* 30 (2002): 42–8.

Bartholomew, Amelia, Sheila Patil, Alastair Mackay, et al. "Baboon Mesenchymal Cells Can Be Genetically Modified to Secrete Human Erythropoietin In Vivo." *Human Gene Therapy* 12, no. 12 (2001): 1527–41.

Bauersachs, Stefan, Susanne E. Ulbrich, Valeri Zakhartchenko, et al. "The Endometrium Responds Differently to Cloned Versus Fertilized Embryos." *Proceedings of the National Academy of Sciences USA* 106, no. 14 (2009): 5681–6.

Baum, C. M., I. L. Weissman, A. S. Tsukamoto, A. M. Buckle, and B. Peault. "Isolation of a Candidate Human Hematopoietic Stem-Cell Population." *Proceedings of the National Academy of Sciences* USA 89, no. 7 (1992): 2804–8.

Baxter Bendus, Allison E., Jacob F. Mayer, Sharon K. Shipley, and William H. Catherino. "Interobserver and Intraobserver Variation in Day 3 Embryo Grading." *Fertility and Sterility* 86, no. 6 (2006): 1608–15.

Baylis F., B. Beagan, J. Johnston, and N. Ram. "Cryopreserved Human Embryos in Canada and Their Availability for Research." *Journal of Obstetrics and Gynaecology Canada* 25, no. 12 (2003): 1026–31.

Baylis, Françoise. "Human Embryonic Stem Cell Lines: The Ethics of Derivation." *Journal of Obstetrics and Gynaecology Canada* 24 (2002): 159–63.

Bearzi, Claudia, Marcello Rota, Toru Hosoda, et al. "Human Cardiac Stem Cells." *Proceedings of the National Academy of Sciences USA* 104, no. 35 (2007): 14068–73.

Beckwith, Francis J. "Defending Abortion Philosophically: A Review of David Boonin's *A Defense of Abortion*." *Journal of Medicine and Philosophy* 31 (2006): 177–203.

Beckwith, Francis J. *Defending Life: A Moral and Legal Case Against Abortion Choice*. Cambridge: Cambridge University Press, 2007.

Begley, Sharon. "From Human Embryos, Hope for 'Spare Parts.'" *Newsweek*, November 16, 1998.

Beltrami, Antonio P., Laura Barlucchi, Daniele Torella, et al. "Adult Cardiac Stem Cells are Multipotent and Support Myocardial Regeneration." *Cell* 114 (2003): 763–76.

Beyhan, Zeki, Pablo J. Ross, Amy E. Iager, et al. "Transcriptional Reprogramming of Somatic Cell Nuclei During Preimplantation Development of Cloned Bovine Embryos." *Developmental Biology* 305, no. 2 (2007): 637–49.

Bieback, Karen and Harald Kluter. "Mesenchymal Stromal Cells from Umbilical Cord Blood." *Current Stem Cell Research & Therapy* 2, no. 4 (2008): 310–23.

Biernaskie, Jeff, Joseph S. Sparling, Jie Liu, et al. "Skin-Derived Precursors Generate Myelinating Schwann Cells that Promote Remyelinating and Functional Recovery after Contusion Spinal Cord Injury." *Journal of Neuroscience* 27, no. 36 (2007): 9545–59.

Binah, Ofer, Katya Dolnikov, Oshra Sadan, et al. "Functional and Developmental Properties of Human Embryonic Stem Cells-Derived Cardiomyocytes." *Journal of Electrocardiology* 40 (2007): S192–6.

Bingham, Richard, ed. *The Works of the Rev. Joseph Bingham*. Vol 6, *The Sixteenth Book of the Antiquities of the Christian Church*. Oxford: Oxford University Press, 1860. http://books.google.com/books?id=gsMPAAAAIAAJ&printsec=frontcover&source=gbs_ge_summary_r&cad=0#v=onepage&q=Lerida&f=false.

Black, Edwin. *War Against the Weak: Eugenics and America's Campaign to Create a Master Race* New York: Four Walls Eight Windows, 2003.

Blight, Andrew R., and William Young. "Central Axons in Injured Cat Spinal Cord Recover Electrophysiological Function Following Remyelination by Schwann Cells." *Journal of the Neurological Sciences* 91 (1989): 15–34.

Blin, Guillaume, David Nury, Sonia Stefanovic, et al. "A Purified Population of Multipotent Cardiovascular Progenitors Derived from Primate Stem Cells Engrafts in Postmyocardial Infarcted Nonhuman Primates." *The Journal of Clinical Investigation* 120, no. 4 (2010): 1125–39.

Bliss, Tonya, Raphael Guzman, Marcel Daadi, and Gary K. Steinberg. "Cell Transplantation Therapy for Stroke." supplement. *Stroke* 38, no S2 (2007): 817–26.

Bodnar, Megan S., Juanito J. Meneses, Ryan T. Rodriguez, and Meri T. Firpo. "Propagation and Maintenance of Undifferentiated Human Embryonic Stem Cells." *Stem Cells and Development* 13, no. 3 (2004): 243–53.

Bolli, Roberto, Atul R. Chugh, Domenico D'Amario, et al. "Cardiac stem cells in patients with ischaemic cardiomyopathy (SCIPIO): initial results of a randomised phase 1 trial." *Lancet* 378, no. 9805 (2011): 1847–57.

Bollini, Sveva, King K. Cheung, Johannes Riegler, et al. "Amniotic Fluid Stem Cells Are Cardioprotective Following Acute Myocardial Infarction." *Stem Cells and Development* 20 (2011): 1985–94.

Bonaros, Nikolaos, Rauend Rauf, Dominik Wolf, et al. "Combined Transplantation of Skeletal Myoblasts and Angiopoietic Progenitor Cells Reduces Infarct Size and Apoptosis and Improves Cardiac Function in Chronic Ischemic Heart Failure." *The Journal of Thoracic and Cardiovascular Surgery* 132, **no.** 6 (2006)**:** 1321–8.

Boonin, David. *A Defense of Abortion*. New York: Cambridge University Press, 2002.

Bortvin, Alex, Kevin Eggan, Helen Skaletsky, et al. "Incomplete Reactivation of *Oct*-4-RelatedGenes in Mouse Embryos Cloned from Somatic Nuclei." *Development* 130 (2003): 1673–80.

Bossolasco, Patrizia, Tiziana Montemurro, Lidia Cova, et al. "Molecular and Phenotypic Characterization of Human Amniotic Fluid Cells and Their Differentiation Potential." *Cell Research* 16 (2006): 329–36.

Brambrink, Tobias, Konrad Hochedlinger, George Bell, and Rudolf Jaenisch. "ES Cells Derived from Cloned and Fertilized Blastocysts are Transcriptionally and Functionally Indistinguishable." *Proceedings of the National Academy of Sciences USA* 103, no. 4 (2006): 933–8.

Bray, Garth M., Maria Paz Villegas-Perez, Manuel Vidal-Sanz, and Albert J. Aguayo. "The Use of Peripheral Nerve Grafts to Enhance Neuronal Survival, Promote Growth and Permit Terminal Reconstructions in the Central Nervous System of Adult Rats." *Journal of Experimental Biology* 132 (1987): 5–19.

Bréart, Gérard, Henrique Barros, Yolande Wagener, and Sabrina Prati. "Characteristics of the Childbearing Population in Europe." supplement, *European Journal of Obstetrics, Gynecology, and Reproductive Biology* 111, no. S1 (2003): S45–52.

Brevini, Tiziana A. L., Georgia Pennarossa, Magda deEguileor, et al. "Parthenogenetic Cell Lines: An Unstable Equilibrium Between Pluripotency and Malignant Transformation." *Current Pharmaceutical Biotechnology* 12 (2011): 206–12.

Breymann, Christian, Dörthe Schmidt, and S. P. Hoerstrup. "Umbilical Cord Cells and a Source of Cardiovascular Tissue Engineering." *Stem Cell Reviews* 2 (2006): 87–92.

Brinton, Louise A., Kamran S. Moghissi, Bert Scoccia, Carolyn L. Westhoff, and Emmet J. Lamb. "Ovulation Induction and Cancer Risk." *Fertility and Sterility* 83, no. 2 (2005):261–74.

Brown, Carolyn. "Stem Cell Tourism Poses Risks." *Canadian Medical Association Journal* 184, no. 2 (2012): E121–2.

Bryant, Peter J., and Philip H. Schwartz. "Stem Cells." In *Fundamentals of the Stem Cell Debate*, edited by Kristen Renwick Monroe, Ronald B. Miller, and Jerome Tobis, 10–36. Berkeley, CA: University of California Press, 2008.

Bunge, Mary Bartlett. "Novel Combination Strategies to Repair the Injured Mammalian Spinal Cord." *Journal of Spinal Cord Medicine* 31, no. 3 (2008): 262–69.

Buratovich, Michael. *Beyond the Dish* (blog). www.beyondthedish.wordpress.com.

———. "May 12, 2006: Biomedical Researcher Hwang Woo-Suk is Charged With Embezzlement and Other Legal Offenses in Faked Research." In *Great Events from History: Modern Scandals*, edited by Carl L. Bankston III, 1094–7. Pasadena, CA: Salem Press, 2009.

———. "New Paper Throws Doubt on the Existence of Very Small Embryonic Stem Cells in Bone Marrow." Beyond the Dish (blog), July 30, 2013. Online: http://beyondthedish.wordpress.com/2013/07/30/new-paper-throws-doubt-on-the-existence-of-very-small-embryonic-stem-cells-in-bone-marrow/.

Burridge, Paul W., Susan Thompson, Michal A. Millrod, et al. "A Universal System for Highly Efficient Cardiac Differentiation of Human Induced Pluripotent Stem Cells that Eliminates Interline Variability." *PloS ONE* 6, no. 4 (2011): e18293. doi:10.1371/journal.pone.001829.

Burt, Richard K., Ann Traynor, Laisvyde Statkute, et al. "Nonmyeloablative Hematopoietic Stem Cell transplantation for Systemic Lupus Erythematosus." *Journal of the American Medical Association* 295, no. 5 (2006): 527–35.

Burt, Richard K., Yvonne Loh, William Pearce, et al. "Clinical Applications of Blood-Derived and Marrow-Derived Stem Cells for Nonmalignant Diseases." *Journal of the American Medical Association* 299 (2008): 925–35.

Busch, Sarah A., and Jerry Silver. "The Role of Extracellular Matrix in CNS Regeneration." *Current Opinion in Neurobiology* 17 (2007): 120–7.

Byrnes, Malcolm W. "The Flawed Scientific Basis of the Altered Nuclear Transfer-Oocyte Assisted Reprogramming (ANT-OAR) Proposal." *Stem Cell Reviews* 3 (2007): 60–65.

Camargo, Luiz Sergio de A., Anne M. Powell, Vicente R. do Vale Filho, and Robert J. Wall. "Comparison of Gene Expression in Individual Preimplantation Bovine Embryos Produced by *In Vitro* Fertilisation or Somatic Cell Nuclear Transfer." *Reproduction, Fertility and Development* 17, no. 5 (2005): 487–96.

Campagnoli, Cesare, Irene A. G. Roberts, Sailesh Kumar, Phillip R. Bennett, Ilaria Bellantuono, and Nicholas M. Fisk. "Identification of Mesenchymal Stem/Progenitor Cells in Human First-Trimester Fetal Blood, Liver and Bone Marrow." *Blood* 98 (2001): 1198–201.

Cao, Feng, Dongdong Sun, Chengxiang Li, et al. "Long-Term Myocardial Functional Improvement after Autologous Bone Marrow Mononuclear Cells Transplantation in Patients with ST-Segment Elevation Myocardial Infarction: 4 Years Follow-Up." *European Heart Journal* 30 (2009): 1986–94.

Cao, Feng, Roger A. Wagner, Kitchener D. Wilson, et al. "Transcriptional and Functional Profiling of Human Embryonic Stem Cell-Derived Cardiomyocytes." *PLOS ONE* 3, no. 10 (2008): e3474(1–12). doi:10.1371/journal.pone.0003474.

Caplan, Arthur L. "Attack of the Anti-Cloners." *The Nation*, June 17, 2002. http:// www .thenation.com/article/attack-anti-cloners.

Caroni, Pico, and Martin E. Schwab. "Oligodendrocyte- and Myelin-Associated Inhibitors of Neurite Growth in the Adult Nervous System." *Advances in Neurology* 61 (1993): 175–9.

Carraro, Gianni, Laura Perin, Sargis Sedrakyan, et al. "Human Amniotic Fluid Stem Cells Can Integrate and Differentiate into Epithelial Lung Lineages." *Stem Cells* 26, no. 11 (2008): 2902–11.

Caspi, Oren, Irit Huber, Izhak Kehat, et al. "Transplantation of Human Embryonic Stem Cell-Derived Cardiomyocytes Improves Myocardial Performance in Infarcted Rat Hearts." *Journal of the American College of Cardiology* 50, no. 19 (2007): 1884–93.

Cattoli, Monica, Andrea Borini, and Maria Atonietta Bonu. "Fate of Stored Embryos: Our 10 Years Experience." supplement, *European Journal of Obstetrics, Gynecology, and Reproductive Biology* 115, no. S1 (2004): S16–8.

Centeno, C. J., J. R. Schultz, M. Cheever, B. Robinson, M. Freeman, and W. Marasco. "Safety and Complications Reporting on the Re-Implantation of Culture-Expanded Mesenchymal Stem Cells Using Autologous Platelet Lysate Technique." *Current Stem Cell Research and Therapy* 5, no. 1 (2010): 81–93.

Centeno, C. J., J. R. Schultz, M. Cheever, et al. "Safety and Complications Reporting Update on the Re-Implantation of Culture-Expanded Mesenchymal Stem Cells Using Autologous Platelet Lysate Technique." *Current Stem Cell Research and Therapy* 6, no. 4 (2011): 368–78.

Centers for Disease Control and Prevention, American Society for Reproductive Medicine, Society for Assisted Reproductive Technology. 2006 *Assisted Reproductive Technology Success Rates: National Summary and Fertility Clinic Reports*, Atlanta: U.S. Department of Health and Human Services, Centers for Disease Control and Prevention, 2008. http://www.cdc.gov/art/PDF/508PDF/2006ART.pdf.

Centers for Disease Control and Prevention, American Society for Reproductive Medicine, Society for Assisted Reproductive Technology. 2008 *Assisted Reproductive Technology Success Rates: National Summary and Fertility Clinic Reports*, Atlanta:

U.S. Department of Health and Human Services, 2010. http://www.cdc.gov/art/ART2008/PDF/ART_2008_Full.pdf.

Chambers, Stuart M., Christopher A. Fasano, Eirini P. Papapetrou, Mark Tomishima, Michel Sadelain, and Lorenz Studer. "Highly Efficient Neural Conversion of Human ES and iPS Cells by Dual Inhibition of SMAD Signaling." *Nature Biotechnology* 27 (2009): 275–80.

Chang, Chia-Ming, Chung-Lan Kao, Yuh-Lih Chang, et al. "Placenta-derived multipotent stem cells induced to differentiate into insulin-positive cells." *Biochemical and Biophysical Research Communication* 357 (2007): 414–20.

Chang, Yu-Jen, Daniel Tzu-bi Shih, Ching-Ping Tseng, Tzu-Bou Hsieh, Don-Ching Lee, and Shiaw-Min Hwang. "Disparate Mesenchyme-Lineage Tendencies in Mesenchymal Stem Cells from Human Bone Marrow and Umbilical Cord Blood." *Stem Cells* 24 (2006): 679–85.

Chavatte-Palmer, Pascale, Y. Heyman, C. Richard, et al. "Clinical, Hormone, and Hematologic Characteristics of Bovine Calves Derived from Nuclei from Somatic Cells." *Biology of Reproduction* 66, no. 6 (2002): 1596–1603.

Chawengsaksophak, Kallayanee, Wim de Graaff, Janet Rossant, Jacqueline Deschamps, and Felix Beck. "Cdx2 is Essential for Axial Elongation in Mouse Development." *Proceedings of the National Academy of Sciences USA* 101, no. 20 (2004): 764–5.

Check, Erika. "Ethicists and Biologists Ponder the Price of Eggs." *Nature* 442 (2006): 606–7.

Checkbiotech (blog). http://checkbiotech.org.

Chen, Alice A., David K. Thomas, Luvena L. Ong, Robert E. Schwartz, Todd R. Golub, and Sangeeta N. Bhatia. "Humanized Mice with Ectopic Artificial Liver Tissues." *Proceedings of the National Academy of Sciences USA* 108, no. 29 (2011): 11842–7.

Chen, Shao-liang, Zhizhong Liu, Nailiang Tian, et al. "Intracoronary Transplantation of Autologous Bone Marrow Mesenchymal Stem Cells for Ischemic Cardiomyopathy Due to Isolated Chronic Occluded Left Anterior Descending Artery." *Journal of Invasive Cardiology* 18, no. 11 (2006): 552–6.

Chen, You-Ren, Yang Li, Li Chen, Xin-Chun Yang, Pi-Xiong Su, and Jun Cai. "The Infarcted Cardiac Microenvironment Cannot Selectively Promote Embryonic Stem Cell Differentiation into Cardiomyocytes." *Cardiovascular Pathology* 20, no. 2 (2011): 77–83.

Cheng, Linzhao, Nancy F. Hansen, Ling Zhao, et al. "Low Incidence of DNA Sequence Variation in Human Induced Pluripotent Stem Cells Generated by Nonintegrating Plasmid Expression." *Cell Stem Cell* 10, no. 3 (2012): 337–44.

Chesné, Patrick, Pierre G. Adenot, Céline Viglietta, Michel Baratte, Laurent Boulanger, and Jean-Paul Renard. "Cloned Rabbits Produced by Nuclear Transfer from Adult Somatic Cells." *Nature Biotechnology* 20 (2002): 366–9.

Chistiakov, Dimitry A. "Endogenous and Exogenous Stem Cells: A Role in Lung Repair and Use in Airway Tissue Engineering and Transplantation." *Journal of Biomedical Science* 17 (2010): 92.

Choi, Kyung-Dal, Junying Yu, Kim Smuga-Otto, et al. "Hematopoietic and Endothelial Differentiation of Human Induced Pluripotent Stem Cells." *Stem Cells* 27, no. 3 (2009): 559–67.

Chung, Young, Irina Klimanskaya, Sandy Becker, et al. "Embryonic and Extraembryonic Stem Cells Lines Derived from Single Mouse Blastomeres." *Nature* 439 (2006): 216–19.

Chung, Young, Irina Klimanskaya, Sandy Becker, et al. "Human Embryonic Stem Cell Lines Generated without Embryo Destruction." *Cell Stem Cell* 2 (2008): 113–7.

Cibelli, Jose B., Keith H. Campbell, George E. Seidel, Michael D. West, and Robert P. Lanza. "The Health Profile of Cloned Animals." *Nature Biotechnology* 20 (2002): 13–14

Cibelli, Jose B., Steve L. Stice, Paul J. Golueke, et al. "Cloned Transgenic Calves Produced from Nonquiescent Fetal Fibroblasts." *Science* 280 (1998): 1256–8.

ClinicalTrials.gov. A Service of the National Institutes of Health. http://clinicaltrials.gov/ct2/home.

Cloutier, Frank, Monica M. Siegenthaler, Gabriel Nistor, and Hans S. Keirstead. "Transplantation of Human Embryonic Stem Cell-Derived Oligodendrocyte Progenitors into Rat Spinal Cord Injuries Does Not Cause Harm." *Regenerative Medicine* 1, no. 4 (2006): 469–79.

Cohen, Jacques, Dagan Wells, and Santiago Munné. "Removal of 2 Cells from Cleavage Stage Embryos is Likely to Reduce the Efficacy of Chromosomal Tests That are Used to Enhance Implantation Rates." *Fertility and Sterility* 87 (2007): 496–503.

Cohen-Tannoudii, Michel and Charles Babinet. "Beyond 'Knock-Out' Mice: New Perspectives for the Programmed Modification of the Mammalian Genome." *Molecular Human Reproduction* 4, no. 10 (1998): 929–38.

Cohn, Jay N. Roberto Ferrari, and Norman Sharpe. "Cardiac Remodeling – Concepts and Clinical Implications: A Consensus Paper from an International Forum on Cardiac Remodeling." *Journal of the American College of Cardiology* 35, no. 3 (2000): 569–82.

Colombo, Alessandro, Massimo Castellani, Emanuela Piccaluga, et al. "Myocardial Blood Flow and Infarct Size after CD133+ Cell Injection in Large Myocardial Infarction with Good Recanalization and Poor Reperfusion: Results from a Randomized Controlled Trial." *Journal of Cardiovascular Medicine* 12, no. 4 (2011): 239–48.

Colombo, Roberto. "Altered Nuclear Transfer as an Alternative Way to Human Embryonic Stem Cells: Biological and Moral Notes." *Communio* 31 (2004): 645–8.

Condic, Maureeen L. "The Basics About Stem Cells." *First Things*, January, 2002. http://www.firstthings.com/article/2007/06/002-the-basics-about-stem-cells-48.

Connery, John. *Abortion: The Development of the Roman Catholic Perspective.* Chicago: Loyola University Press, 1977.

Cortes, Jose Luis, Guillermo Antiñolo, Luis Martínez, et al. "Spanish Stem Cell Bank Interviews Examine the Interest of Couples in Donating Surplus Human IVF Embryos for Stem Cell Research." *Cell Stem Cell* 1 (2007): 17–20.

Cowan, Chad A., Irina Klimanskaya, Jill McMahon, et al. "Derivation of Embryonic Stem-Cell Lines from Human Blastocysts." *New England Journal of Medicine* 350, no. 13 (2004): 1353–6.

Crisan, Mihaela, Solomon Yap, Louis Casteilla, et al. "A Perivascular Origin for Mesenchymal Stem Cells in Multiple Human Organs." *Cell Stem Cell* 3, no. 3 (2008): 301–13.

Cristante, A. F., T. E. P. Barros-Filho, N. Tatsui, et al. "Stem Cells in the Treatment of Chronic Spinal Cord Injury: Evaluation of Somatosensitive Evoked Potentials in 39 Patients." *Spinal Cord* 47 (2009): 733–48.

Cross, J. C., D. Baczyk, N. Dobric, et al. "Genes, Development and Evolution of the Placenta." *Placenta* 24 (2003): 123–30.

Crossway staff. "Scott Klusendorf on Embryonic Stem Cell Research." *Crossway Blog.* http:// www.crossway.org/blog/2009/03/scott-klusendorf-embryonic-stem-cell-research.

Crovace, A., L. Lacitignola, R. deSiena, G. Rossi, and E. Francioso. "Cell Therapy for Tendon Repair in Horses: An Experimental Study." supplement, *Veterinary Research Communications* 31, no. S1, (2007): 281–3.

Cui, Xiaoiun, Haijie Wang, Haidong Guo, Cun Wang, Hong Ao, Xaiogin Liu, and Yu-Zhen Tan. "Transplantation of Mesenchymal Stem Cells Preconditioned with Diazoxide, a Mitochondrial ATP-Sensitive Potassium Channel Opener, Promotes Repair of Myocardial Infarction in Rats." *Tohoku Journal of Experimental Medicine* 220 (2010): 139–47.

da Silva, Lindolfo, Pedro Cesar Chagastelles, and Nance Beyer Nardi. "Mesenchymal Stem Cells Reside in Virtually All Post-Natal Organs and Tissues." *Journal of Cell Science* 119 (2006): 2204–13.

Dale, Alfred William Winterslow. *The Synod of Elvira and Christian Life in the Fourth Century*. London: Macmillan and Co., 1882. http://archive.org/stream/synodofelvirachroodale#page/n6/mode/1up.

Dasari, Venkata Ramesh, Daniel G. Spomar, Christopher S. Gondi, et al. "Axonal Remyelination by Cord Blood Stem Cells after Spinal Cord Injury." *Journal of Neurotrauma* 24, no. 2 (2007): 391–410.

Davis, Darryl R., and Duncan J. Stewart. "Autologous Cell Therapy for Cardiac Repair." *Expert Opinion on Biological Therapy* 11, no. 4 (2011): 489–508.

Davis, Darryl R., Rachel Ruckdeschel Smith, and Eduardo Marbán. "Human Cardiospheres Are a Source of Stem Cells with Cardiomyogenic Potential." *Stem Cells* 28 (2010): 903–4.

De Coppi, Paolo, Georg Bartsch Jr., M. Minhaj Siddiqui, et al. "Isolation of Amniotic Stem Cell Lines with Potential for Therapy." *Nature Biotechnology* 25, no. 1 (2007): 100–6.

de Lacey, Sheryl. "Parent Identity and "Virtual" Children: Why Patients Discard Rather than Donate Unused Embryos." *Human Reproduction* 20, no. 6 (2005): 1661–9.

De Sousa, Paul A., Tim King, Linda Harkness, Lorraine E. Young, Simon K. Walker, and Ian Wilmut. "Evaluation of Gestational Deficiencies in Cloned Sheep Fetuses and Placentae." *Biology of Reproduction* 65 (2001): 23–30.

De Vos, Anick., Catherine Stassen, Martine De Rycke, et al., "Impact of cleavage-stage embryo biopsy in view of PGD on human blastocyst implantation: a prospective cohort of single embryo transfers." *Human Reproduction* 24 (2009): 2988–96.

Dekel, Benjamin, Tatyana Burakova, Fabian D. Arditti, et al. "Human and Porcine Early Kidney Precursors as a New Source for Transplantation." *Nature Medicine* 9 (2003): 53–60.

Delo, Dawn M., John Olson, Pedro M. Baptista, Ralph B. D'Agostino Jr., Anthony Atala, Jian-Ming Zhu, and Shay Soker. "Non-invasive Longitudinal Tracking of Human Amniotic Fluid Stem Cells in the Mouse Heart." *Stem Cells and Development* 17, no. 6 (2008): 1185–94.

Denning, Chris, S. Burl, A. Ainslie et al. "Deletion of the α(1,3) Galalctosyl Transferase (GGTA1) Gene and the Prion Protein (PrP) Gene in Sheep." *Nature Biotechnology* 19 (2001): 559–62.

Dennis, Carina. "Developmental Biology: Synthetic Sex Cells." *Nature* 64 (2003): 364–6.

Deveault, Catherine, Jian Hua Qian, Wafaa Chebaro, et al. "*NLRP7* Mutations in Women with Diploid Androgenetic and Triploid Moles: A Proposed Mechanism for Mole Formation." *Human Molecular Genetics* 18, no. 5 (2009): 888–97.

Devine, Philip. *The Ethics of Homicide.* Ithaca, New York; Cornell University, 1978.

Devine, Steven M., Carrington Cobbs, Matt Jennings, Amelia Bartholomew, and Ron Hoffman. "Mesenchymal Stem Cells Distribute to a Wider Range of Tissues Following Systemic Infusion into Nonhuman Primates." *Blood* 101, no. 8 (2003): 2999–3001.

Devine, Steven M., S. Peter, B. J. Martin, F. Barry, and K. R. McIntosh. "Mesenchymal Stem Cells: Stealth and Suppression." supplement, *Cancer Journal* no. S7 (2001): S76–S82.

Devriendt, Koen. "Hydatidiform Mole and Triploidy: The Role of Genomic Imprinting in Placental Development." *Human Reproduction Update*, 11 (2005): 137–42.

Di Nicola, Massimo, Carmelo Carlo-Stella, Michele Magni, et al. "Human Bone Marrow Stromal Cells Suppress T-Lymphocyte Proliferation Induced by Cellular or Nonspecific Mitogenic Stimuli." *Blood* 99, no. 10 (2002): 3838–43.

Dickenson, Donna. "Commodification of Human Tissue: Implications for Feminist and Developmental Ethics." *Developing World Bioethics* 2, no. 1 (2002): 55–63.

Dimos, John T., Kit T. Rodolfa, Kathy K. Niakan, et al. "Induced Pluripotent Stem Cell Generated from Patients with ALS Can Be Differentiated into Motor Neurons." *Science* 321 (2008): 1218–21.

Ditadi, Andrea, Paolo de Coppi, Olivier Picone, et al. "Human and Murine Amniotic Fluid c-Kit+Lin− Cells Display Hematopoietic Activity." *Blood* 113, no. 17 (2009): 3953–60.

Do No Harm. http://www.stemcellresearch.org.

Dobson, Anthony T., Rajiv Raja, Michael J. Abeyta, Theresa Taylor, Shehua Shen, Christopher Haqq, and Renee A. Reijo Pera. "The Unique Transcriptome Through Day 3 of Human Preimplantation Development." *Human Molecular Genetics* 13, no. 14 (2004): 1461–70.

Doetschman, Thomas C., Harald Eistetter, Margot Katz, Werner Schmidt, and Rolf Kemler. "The *In Vitro* Development of Blastocyst-Derived Embryonic Stem Cell Lines: Formation of Visceral Yolk Sac, Blood Islands and Myocardium." *Journal of Embryology and Experiential Morphology* 87 (1985): 27–45.

Dogru, Murat, and Kazuo Tsubota. "Current Concepts in Ocular Surface Reconstruction." *Seminars in Ophthalmology* 20 (2005): 75–93.

Dohmann, Hans F. R., Emerson C. Perin, Christina M. Takiya, et al. "Transendocardial Autologous Bone Marrow Mononuclear Cell Injection in Ischemic Heart Failure: Postmortem Anatomicopathologic and Immunohistochemical Findings." *Circulation* 112 (2005): 521–6.

Dominici, M., K. Le Blanc, I. Mueller, et al. "Minimal Criteria for Defining Multipotent Mesenchymal Stromal Cells. The International Society for Cellular Therapy Position Statement." *Cytotherapy* 8, no. 4 (2006): 315–7.

Doucette, Ronald. "Olfactory Ensheathing Cells: Potential for Glial Cell Transplantation into Areas of CNS Injury." *Current Pharmacology and Biotechnology* 5 (2004): 567–84.

Dowell, Joshua D., Michael Rubart, Kishore B. S. Pasumarthi, Mark H. Soonpaa, and Loren J. Field. "Myocyte and Myogenic Stem Cell Transplantation in the Heart."*Cardiovascular Research* 58, no. 2 (2003): 336–50.

Drukker, Micha, Gil Katz, Achia Urbach, et al. "Characterization of the Expression of MHC Proteins in Human Embryonic Stem Cells." *Proceedings of the National Academy of Sciences USA* 99 (2002): 9864–9.

Drukker, Micha, Helena Katchman, Gil Katz, et al. "Human Embryonic Stem Cells and their Derivatives are Less Susceptible to Immune Rejection than Adult Cells." *Stem Cells* 24 (2006): 221–9.

Duckers H., J. Houtgraaf, C. Hehrlein, et al. "Final results of a Phase IIa, Randomised, Open-Label Trial to Evaluate the Percutaneous Intramyocardial Transplantation of Autologous Skeletal Myoblasts in Congestive Heart Failure Patients: the SEISMIC Trial." *Eurointervention* 6, no. 7 (2011): 805–12.

Duncan, Francesca E., Paula Stein, Carmen J. Williams, and Richard M. Schultz. "The Effect of Blastomere Biopsy on Preimplantation Mouse Embryo Development and Global Gene Expression." *Fertility and Sterility* 91 (2009): 1462–5.

Dunstan, Gordon Reginald. "The Human Embryo in the Western Moral Tradition." In *The Status of the Embryo; Perspectives from Moral Tradition*, edited by Gordon Reginald Dunstan and Mary J. Seller, 39ff. London: King Edward's Hospital Fund, 1988.

Ebert, Allison D., Junying Yu, Ferrill F. Rose, Jr., et al. "Induced Pluripotent Stem Cells from a Spinal Muscular Atrophy Patient." *Nature* 457 (2009): 277–81.

Edwards, Robert G., S. B. Fishel, J. Cohen, et al. "Factors Influencing the Success of In Vitro Fertilization for Alleviating Human Infertility." *Journal of In Vitro Fertilization and Embryo Transfer* 1 (1984): 752–68.

Eftekharpour, Eftekhar, Soheila Karimi-Abdolrezaee, Jian Wang, Hossam El Beheiry, Cindi Morshead, and Michael G. Fehlings. "Myelination of Congenitally Dysmyelinated Spinal Cord Axons by Adult Neural Precursor Cells Results in Formation of Nodes of Ranvier and Improved Axonal Conduction." *Journal of Neuroscience* 27, no. 13 (2007): 3416–28.

Epistle of Barnabas. Translated by Alexander Roberts and James Donaldson. Amazon Digital Services, 2011. Kindle edition.

Eppig, John J., and Marilyn J. O'Brien. "Development In Vitro of Mouse Oocytes from Primordial Follicles." *Biology of Reproduction* 54 (1996): 197–207.

Erices, Alejandro A., C. I. Allers, P. A. Conget, C. V. Rojas, and J. J. Minguell. "Human Cord Blood-Derived Mesenchymal Stem Cells Home and Survive in the Marrow of Immunodeficient Mice After Systemic Infusion." *Cell Transplant* 12 (2003): 555–61.

Estrela, Carlos, Ana Helena Gonçalves de Alencar, Gregory Thomas Kitten, Eneida Franco Vencio, and Elisandra Gava. "Mesenchymal Stem Cells in the Dental Tissues: Perspectives for Tissue Regeneration." *Brazilian Dental Journal* 22, no. 2 (2011): 91–8.

Ethics and Policy Center. "Production of Pluripotent Stem Cells by Oocyte Assisted Reprogramming: Joint Statement." June 20, 2005. http://www.eppc.org/publications/pubID.2374/pub_detail.asp.

Ethics Committee of the American Society for Reproductive Medicine. "Financial Incentives in Recruitment of Oocyte Donors." supplement, *Fertility and Sterility* 82, no. S1 (2004): S240–4.

Evans, Martin, J., and Matthew H. Kaufman. "Establishment in Culture of Pluripotential Cells from Mouse Embryos." *Nature* 292 (1981): 154–6.

Falzacappa, Maria V. Verga, Chiara Ronchini, Linsey B. Reavie, and Pier G. Pelicci. "Regulation of self-renewal in normal and cancer stem cells." *FEBS Journal* 279 (2012): 3559–72.

Farin, Charolette. E., Peter W. Farin, and Jorge A. Piedrahita. "Development of Fetuses from In Vitro–Produced and Cloned Bovine Embryos." *Journal of Animal Science* 82 (2004): E53–62.

Farin, Peter W., Jorge A. Piedrahita, and Charlotte E. Farin. "Errors in Development of Fetuses and Placentas from In Vitro-Produced Bovine Embryos." *Theriogenology* 65, no. 1 (2006): 178–91.

Fauser, Bart C. M. J., Paul Devroey, and Nick Macklon. "Multiple Birth Resulting from Ovarian Stimulation for Subfertility Treatment." *Lancet* 365 (2005): 1807–16.

Fazel, Shafie, Massimo Cimini, Liwen Chen, et al. "Cardioprotective c-kit+ Cells are from the Bone Marrow and Regulate the Myocardial Balance of Angiogenic Cytokines." *Journal of Clinical Investigation* 116, no. 7 (2006): 1865–77.

Feldman, David M. *Marital Relations: Birth Control and Abortion in Jewish Law.* New York: Schocken Books, 1974.

Féron, F., C. Perry, J. Cochrane, et al. "Autologous Olfactory Ensheathing Cell Transplantation in Human Spinal Cord Injury." *Brain* 128 (2005): 2951–60.

Ferrand, Jonathan, Danièle Noël, Philippe Lehours, et al. "Human Bone Marrow-Derived Stem Cells Acquire Epithelial Characteristics through Fusion with Gastrointestinal Epithelial Cells." *PLoS One*. 6, no. 5 (2011): e19569. doi:10.1371/journal.pone .0019569.

Ferrari, Giuliana, Gabriella Cusella–De Angelis, Marcello Coletta, et al. "Muscle Regeneration by Bone Marrow-Derived Myogenic Progenitors." *Science* 279 (1998): 1528–30.

Field, Loren J. "Myocyte and Myogenic Stem Cell Transplantation in the Heart." *Circulation Research* 58 (2003): 336–50.

Fineschi, Vittorio, Margherita Neri, Sabina Di Donato, Cristoforo Pomara, Irene Riezzo, and Emanuela Turillazzi. "An Immunohistochemical Study in a Fatality Due to Ovarian Hyperstimulation Syndrome." *International Journal of Legal Medicine* 120, no. 5 (2006): 293–9.

Fischer, Joannie. "The First Clone." *U.S. News and World Report,* November 25, 2001. http:// www.usnews.com/usnews/culture/articles/011203/archive_019784.htm.

Flemming, Tom P., Bhavwanti Sheth, and Irina Fesenko. "Cell Adhesion in the Preimplantation Embryo and Its Role in Trophectoderm Differentiation and Blastocyst Morphogenesis." *Frontiers in Bioscience* 6 (2001): 1000–1007.

Flores-Ramírez, Ramiro, Artemio Uribe-Longoria, María M. Rangel-Fuentes, et al. "Intracoronary Infusion of CD133+ Endothelial Progenitor Cells Improves Heart Function and Quality of Life in Patients with Chronic Post-Infarct Heart Insufficiency." *Cardiovascular Revascularization Medicine* 11, no. 2 (2010): 72–8.

Földes, Gábor, Sian Harding, and Nadire N Ali. "Cardiomyocytes from Embryonic Stem Cells: Towards Human Therapy." *Expert Opinion on Biological Therapies* 8, no. 10 (2008): 1473–83.

Ford, Norman M. *When Did I Begin?: Conception of the Human Individual in History, Philosophy and Science*. Cambridge, UK: Cambridge University Press, 1991.

Fouillard, L., M. Bensidhoum, D. Bories, et al. "Engraftment of Allogeneic Mesenchymal Stem Cells in the Bone Marrow of a Patient with Severe Idiopathic Aplastic Anemia Improves Stroma." *Leukemia* 17 (2003): 474–6.

Fox, Stuart Ira. *Human Physiology*. 10th ed. New York: McGraw-Hill, 2008.

Frangoul, Haydar and Jennifer Domm. "The Quality of Privately Banked Cord Blood." *Pediatric Blood & Cancer* 57 (2011): 1248.

Frantz, Simon. "Embryonic Stem Cell Pioneer Geron Exits Field, Cuts Losses." *Nature Biotechnology* 30 (2012): 12–13.

Freed, Curt R., Maureen A. Leehey, Michael Zawada, Kimberly Bjugstad, Laetitia Thompson and Robert E. Breeze. "Do Patients with Parkinson's Disease Benefit from Embryonic Dopamine Cell Transplantation?", supplement, *Journal of Neurology* 250, no. S3 (2003): III44–6.

Freed, Curt R., Paul E. Greene, Robert E. Breeze, et al. "Transplantation of Embryonic Dopamine Neurons for Severe Parkinson's Disease." *The New England Journal of Medicine* 344, no. 10 (2001): 710–9.

French, Andrew J., Catharine A. Adams, Linda S. Anderson, John R. Kitchen, Marcus R. Hughes, and Samuel H. Wood. "Development of Human Cloned Blastocysts Following Somatic Cell Nuclear Transfer with Adult Fibroblasts." *Stem Cells* 26, no. 2 (2008): 485–93.

Fumento, Michael. "Short on Facts: The False Argument over Embryonic-Stem-Cell Research." *National Review Online,* July 23, 2001. http://old.nationalreview.com/comment/comment-fumento072301.shtml.

Fujimoto, Yusuke, Masahiko Abematsu, Anna Falk, et al. "Treatment of a Mouse Model of Spinal Cord Injury by Transplantation of Human iPS Cell-Derived Long-term Self-renewing Neuroepithelial-like Stem Cells." *Stem Cells* (2012): doi: 10.1002/stem.1083.

Fundele, Reinald, Michael L. Norris, Sheila C. Barton, Wolf Reik, and M. Azim Surani. "Systematic Elimination of Parthenogenetic Cells in Mouse Chimeras." *Development* 106 (1989): 29–35.

Fundele, Reinald, Michael L. Norris, Sheila C. Barton, et al. "Temporal and Spatial Selection Against Parthenogenetic Cells During Development of Fetal Chimeras." *Development* 108 (1990): 203–11.

Galli, Cesare, Irina Lagutina, Gabriella Crotti, et al. "Pregnancy: A Cloned Horse Born to Its Dam Twin." *Nature* 424 (2003): 635.

Galli, Cesare, Roberto Duchi, Robert M. Moor, and Giovanna Lazzari. "Mammalian Leukocytes Contain All the Genetic Information Necessary for the Development of a New Individual." *Cloning* 1 (1999): 161–70.

Garderet, Laurent, Nicolas Dulphy, Corinne Douay, et al. "The Umbilical Cord Blood αβ T-Cell Repertoire: Characteristics of a Polyclonal and Naive but Completely Formed Repertoire." *Blood* 91, no. 1 (1998): 340–6.

Gardner, David K., Michelle Lane, John Stevens, Terry Schlenker, and William B. Schoolcraft. "Blastocyst Score Affects Implantation and Pregnancy Outcome: Towards a Single Blastocyst Transfer." *Fertility and Sterility* 73, no. 6 (2000): 1155–8.

Garger, Ilya. "Take Heart." *Time Magazine*, November 12, 2005. http:// www.time.com/time/magazine/article/0,9171,1129470,00.html.

Gates, Charley B., Tharun Karthikeyan, Freddie Fu, and Johnny Huard. "Regenerative Medicine for the Musculoskeletal System Based on Muscle-Derived Stem Cells." *Journal of the American Academy of Orthopaedic Surgeons* 16, no. 2 (2008): 68–76.

Geoffrey, Robert, Liselotte Mettler, and D. E. Walters. "Identical Twins and In Vitro Fertilization." *Journal of In Vitro Fertilization and Embryo Transfer* 3 (1986): 114–7.

George, Robert P., and Christopher Tollefsen. *Embryo: A Defense of Human Life.* New York: Doubleday, 2008.

George, Robert P., and Patrick Lee. "Acorns and Embryos." *The New Atlantis* 7 (Fall 2004 / Winter 2005): 90–100.

Gluckman, Elaine and Vanderson Rocha. "Cord Blood Transplantation: State of the Art." *Haematologica* 94, no. 4 (2009): 451–4.

Gluckman, Eliane, Hal E. Broxmeyer, Arleen D. Auerbach, et al. "Hematopoietic Reconstitution in a Patient with Fanconi's Anemia by Means of Umbilical-Cord Blood from an HLA-Identical Sibling." *New England Journal of Medicine* 321 (1989): 1174–8.

Glud, Eva, Susanne Kriiger Kjaer, Rebecca Troisi, and Louise A. Brinton. "Fertility Drugs and Ovarian Cancers." *Epidemiological Reviews* 20, no. 2 (1998): 237–57.

Goldberg, Jeffrey M., Tommaso Falcone, and Marjan Attaran. "In Vitro Fertilization Update." *Cleveland Clinic Journal of Medicine* 74, no. 5 (2007): 329–38.

González, Federico, Stéphanie Boué, and Juan Carlos Izpisúa Belmonte. "Methods for Making Induced Pluripotent Stem Cells: Reprogramming à la Carte." *Nature Reviews Genetics* 12 (2011): 231–42.

Goossens, Veerle, Martine De Rycke, Anick De Vos, et al. "Diagnostic Efficiency, Embryonic Development and Clinical Outcome After the Biopsy of One or Two Blastomeres for Preimplantation Genetic Diagnosis." *Human Reproduction* 23 (2008): 481–92.

Gore, Athurva, Zhe Li, Ho-Lim Fung, et al. "Somatic Coding Mutations in Human Induced Pluripotent Stem Cells." *Nature* 471 (2011): 63–7.

Gorman, Michael J. *Abortion and the Early Church: Christian, Jewish, and Pagan Attitudes in the Greco-Roman World.* Eugene, Oregon: Wipf and Stock Publishers, 1982.

Grauss, Robert W., Elizabeth M. Winter, John van Tuyn, et al. "Mesenchymal Stem Cells from Ischemic Heart Disease Patients Improve Left Ventricular Function after Acute Myocardial Infarction." *American Journal of Physiology – Heart and Circulatory Physiology* 293 (2007): H2438–47.

Grauss, Robert W., John van Tuyn, Paul Steendijk, et al. "Forced Myocardin Expression Enhances the Therapeutic Effect of Human Mesenchymal Stem Cells after Transplantation in Ischemic Mouse Hearts." *Stem Cells* 26, **no.** 4 (2008)**:** 1083–93.

Green, Andrew J., David E. Barton, Peter Jenks, J. Pearson, and Y. R. Yates. "Chimaerism Shown by Cytogenetics and DNA Polymorphism Analysis." *Journal of Medical Genetics* 31 (1994): 816–7.

Green, Ronald M. "Can We Develop Ethically Universal Embryonic Stem-Cell Lines?" *Nature Reviews Genetics* 8 (2007): 480–5.

Green, Ronald M. *The Human Embryo Research Debates: Bioethics in the Vortex of Controversy.* New York: Oxford University Press, 2001.

Gregory IX (pope). *Quibque Libri Decretalium.* Edited by Raymond Peñafort, AD 1234. http://www.hs-augsburg.de/~harsch/Chronologia/Lspost13/GregoriusIX/gre_0000.html.

Griffin, Chad and Noelle Gambill. "Somatic Cell Nuclear Transfer Gives Old Animals Youthful Immune Cells." *Advanced Cell Technology* (press release). June 29, 2005. http://www.advancedcell.com/documents/0000/0147/B_17_Somatic_Cell_Nuclear_Transfer_Gives_Old_Animals_Youthful_Immune_Cells_6_29_05.pdf.

Grigoriadis, Nikolaos, Tamir Ben-Hur, Dimitrios Karussis, and Ioannis Milonas. "Axonal Damage in Multiple Sclerosis: A Complex Issue in a Complex Disease." *Clinical Neurology and Neurosurgery* 106, no. 3 (2004): 211–7.

Gronthos, Stan, Dawn M. Franklin, Holly A. Leddy, Pamela G. Robey, Robert W. Storms, and Jeffrey M. Gimble. "Surface Protein Characterization of Human Adipose Tissue-Derived Stromal Cells." *Journal of Cell Physiology* 189 (2001): 54–63.

Gruen, Lori, and Laura Grabel. "Concise Review: Scientific and Ethical Roadblocks to Human Embryonic Stem Cell Therapy." *Stem Cells* 24 (2006): 2162–9.

Gruen, Lori. "Oocytes for Sale?" In *Stem Cell Research: The Ethical Issues*, edited by Lori Gruen, Laura Grabel, and Peter Singer, 143–67. Malden, MA: Blackwell Publishing, 2007.

———. "Oocytes for Sale?" *Metaphilosophy* 38, no. 2 (2007): 285–308.

Guerif, Fabrice, Rachel Bidault, Veronique Cadoret, Marie-Laure Couet, Jacques Lansac, and Dominique Royere. "Parameters Guiding Selection of Best Embryos for Transfer after Cryopreservation: A Reappraisal." *Human Reproduction* 17, no. 5 (2002): 1321–6.

Guimaraes-Souza, Nadia K., Liliya M. Yamaleyeva, Tamer AbouShwareb, Anthony Atala, and James J. Yoo. "*In vitro* Reconstitution of Human Kidney Structures for Renal Cell Therapy." *Nephrology, Dialysis, Transplantation* (2012): doi: 10.1093/ndt/gfr785.

Gyöngyösi, M., A. Posa, N. Pavo, et al. "Differential Effect of Ischaemic Preconditioning on Mobilisation and Recruitment of Haematopoietic and Mesenchymal Stem Cells in Porcine Myocardial Ischaemia-Reperfusion." *Thrombosis and Haemostasis* 104 (2010): 376–84.

Hacein-Bey-Abina, S., C. Von Kalle, M. Schmidt, et al. "LMO2-Associated Clonal T Cell Proliferation in Two Patients after Gene Therapy for SCID-X1." *Science* 302, no. 5644 (2003): 415–9.

Haider, Husnain Kh., Ye Lei, and Muhammad Ashraf. "MyoCell, a Cell-Based Autologous Skeletal Myoblast Therapy or the Treatment of Cardiovascular Diseases." *Current Opinion in Molecular Therapy* 10, no. 6 (2008): 611–21.

Hall, Judith G. "Twinning." *Lancet* 362 (2003): 735–43.

Hall, Vanessa J., Petra Stojkovic, and Miodrag Stojkovic. "Using Therapeutic Cloning to Fight Human Disease: A Conundrum or Reality?" *Stem Cells* 24 (2006): 1628–37.

Han, Steve S. W., Ying Liu, Carla Tyler-Polsz, Mahendra S. Rao, and Itzhak Fischer. "Transplantation of Glial-Restricted Precursor Cells into the Adult Spinal Cord: Survival, Glial-Specific Differentiation, and Preferential Migration in White Matter." *Glia* 45 (2004): 1–16.

Hanna, Jacob, Marius Wernig, Styliani Markoulaki, et al. "Treatment of Sickle Cell Anemia Mouse Model with iPS Cells Generated from Autologous Skin." *Science* 318 (2007): 1920–3.

Hansis, Christoph and Robert G. Edwards. "Initial Differentiation of Blastomeres in 4-Cell Human Embryos and Its Significance for Early Embryogenesis and Implantation." *Reproductive BioMedicine Online* 11 (2005): 206–18.

Hansis, Christoph, James A. Grifo, Lewis C. Krey. "Candidate Lineage Marker Genes in Human Preimplantation Embryos." *Reproductive BioMedicine Online* 8, (2004): 577–83.

Hare, Joshua M., Jay H. Traverse, Timothy D. Henry, et al. "A Randomized, Double-Blind, Placebo-Controlled, Dose-Escalation Study of Intravenous Adult Human Mesenchymal Stem Cells (Prochymal) after Acute Myocardial Infarction." *Journal of the American College of Cardiology* 54, no. 24 (2009): 2277–86.

Hartwell, Leland H., Leroy Hood, Michael Goldberg, Ann E. Reynolds, Lee M. Silver, and Ruth C. Veres. *Genetics: From Genes to Genomes*. New York: McGraw-Hill, 2008.

Hatano, Shin-ya, Masako Tada, Hironobu Kimura, et al. "Pluripotential Competence of Cells Associated with *Nanog* Activity." *Mechanisms of Development* 122, no. 1 (2005): 67–79.

Hatzistergos, Konstantinos E., Henry Quevedo, Behzad N. Oskouei, et al. "Bone Marrow Mesenchymal Stem Cells Stimulate Cardiac Stem Cell Proliferation and Differentiation." *Circulation Research* 107, no. 7 (2010): 913–22

Hauser, Peter V., Roberta De Fazio, Stefania Bruno, Simona Sdei, Cristina Grange, Benedetta Bussolati, Chiara Benedetto, and Giovanni Camussi. "Stem Cells Derived from Human Amniotic Fluid Contribute to Acute Kidney Injury Recovery." *American Journal of Pathology* 177 (2010): 2011–21.

Hauzman, E. E., and Z. Papp. "Conception without the Development of a Human Being." *Journal of Perinatal Medicine* 36, no. 2 (2008): 175–7.

Hays, Kristen. "Cloned Cat Isn't A Carbon Copy." *CBS News Tech.* February 11, 2009. http://www.cbsnews.com/stories/2003/01/21/tech/main537380.shtml.

Hayward, Bruce E., Michel De Vos, and Nargese Talati. "Genetic and Epigenetic Analysis of Recurrent Hydatidiform Mole." *Human Mutation* 30 (2009): E629–39.

He, Jia-Qiang, Yue Ma, Youngsook Lee, James A. Thomson, and Timothy J. Kamp. "Human Embryonic Stem Cells Develop into Multiple Types of Cardiac Myocytes." *Circulation Research* 93 (2003): 31–9.

He, Kun-Lun, Geng-Hua Yi, Warren Sherman, et al. "Autologous Skeletal Myoblast Transplantation Improved Hemodynamics and Left Ventricular Function in Chronic Heart Failure Dogs." *Journal of Heart and Lung Transplantation* 24, no. 11 (2005): 1940–9.

He, Qing, Pedro T. Trindade, Michael Stumm, et al. "Fate of Undifferentiated Mouse Embryonic Stem Cells within the Rat Heart: Role of Myocardial Infarction and Immune Suppression." *Journal of Cellular and Molecular Medicine* 13, no. 1 (2009): 188–201.

Heile, Anna M. B., Christine Wallrapp, Petra M. Klinge, et al. "Cerebral Transplantation of Encapsulated Mesenchymal Stem Cells Improves Cellular Pathology after Experimental Traumatic Brain Damage." *Neuroscience Letters* 463 (2009): 176–81.

Hejčl, Ales, Lucie Urdzikova, Jiri Sedy, et al. "Acute and Delayed Implantation of Positively Charged 2-Hydroxyethyl Methacrylate Scaffolds in Spinal Cord Injury in the Rat." *Journal of Neurosurgery: Spine* 8, no. 1 (2008): 67–73.

Hematti, Peiman. "Role of Mesenchymal Stromal Cells in Solid Organ Transplantation." *Transplantation Reviews* 22, no. 4 (2008): 262–73.

Hierlihy, Andrée M., Patrick Seale, Corrinne G. Lobe, Michael A. Rudnicki, and Lynn A. Megeney. "The Post-Natal Heart Contains a Myocardial Stem Cell Population." *FEBS Letters* 530, no. 1 (2002): 239–43.

Hill, Jonathan R., Robert C. Burghardt, Karen Jones, et al. "Evidence for Placental Abnormality as the Major Cause of Mortality in First-Trimester Somatic Cell Cloned Bovine Fetuses." *Biology of Reproduction* 63 (2000): 1787–94.

Hill, Mark. *UNSW Embryology*. http://php.med.unsw.edu.au/embryology/index.php?title=Main_Page.

Hjelm, Brooke E., Jon B. Rosenberg, Szabolcs Szelinger, et al. "Induction of Pluripotent Stem Cells from Autopsy Donor-Derived Somatic Cells." *Neuroscience Letters* 502, no. 3 (2011): 219–24.

Hodgson, Denice M., Atta Behfar, Leonid V. Zingman, et al. "Stable Benefit of Embryonic Stem Cell Therapy in Myocardial Infarction." *American Journal of Physiology: Heart and Circulation Physiology* 287 (2004): H471–9.

Hoffman, David I., Gail L. Zellman, C. Christine Fair, et al. "Cryopreserved Embryos in the United States and Their Availability for Research." *Fertility and Sterility* 79 (2003): 1063–9.

Hombach-Klonisch, Sabine, Soumya Panigrahi, Iran Rashedi, et al. "Adult Stem Cells and Their Trans-differentiation Potential—Perspectives and Therapeutic Applications." *Journal of Molecular Medicine* 86 (2008): 1301–14.

Honmou, Osamu, Paul A. Felts, Stephen G. Waxman, and Jeffery D. Kocsis. "Restoration of Normal Conduction Properties in Demyelinated Spinal Cord Axons in the Adult Rat by Transplantation of Exogenous Schwann Cells." *Journal of Neuroscience* 16, no. 10 (1996): 3199–208.

Hoogduijn, Martin J., M. J. Crop, A. M. A. Peeters, et al. "Donor-Derived Mesenchymal Stem Cells Remain Present and Functional in the Transplanted Human Heart." *American Journal of Transplantation* 9, no. 1 (2009): 222–30.

Hook, Christopher C. "In Vitro Fertilization and Stem Cell Harvesting from Human Embryos: The Law and Practice in the United States." *Polskie Archiwum Medycyny Wewnętrznej* 128, no. 7–8 (2010): 282–9.

Horwitz, Edwin M., Darwin J. Prockop, Lorraine A. Fitzpatrick, et al. "Transplantability and Therapeutic Effects of Bone Marrow-Derived Mesenchymal Cells in Children with Osteogenesis Imperfecta." *Nature Medicine* 5 (1999): 309–13.

Horwitz, Edwin M., Darwin J. Prockop, Patricia L. Gordon, et al. "Clinical Responses to Bone Marrow Transplantation in Children with Severe Osteogenesis Imperfecta." *Blood* 97 (2001): 1227–31.

Hou, Mai, Ke-ming Yang, Hao Zhang, et al. "Transplantation of Mesenchymal Stem Cells from Human Bone Marrow Improves Damaged Heart Function in Rats." *International Journal of Cardiology* 115, no. 2 (2007): 220–8.

Howard, Michael J., Su Liu, Frank Schottler, B. Joy Snider, and Mark F. Jacquin. "Transplantation of Apoptosis-Resistant Embryonic Stem Cells into the Injured Rat Spinal Cord." *Somatosensory and Motor Research* 22, no. 3 (2005): 37–44.

Huang, Hongyun, Hongmei Wang, Lin Chen, et al. "Influence Factors for Functional Improvement after Olfactory Ensheathing Cell Transplantation for Chronic Spinal Cord Injury." *Zhongguo Xiu Fu Chong Jian Wai Ke Za Zhi* 20, no. 4 (2006): 434–8.

Huang, Hongyun, Hongmei Wang, Lin Chen, et al. "Influence of Patients' Age on Functional Recovery after Transplantation of Olfactory Ensheathing Cells into Injured Spinal Cord Injury." *Chinese Medical Journal* 116 (2003): 1488–91.

Huang, Hongyun, Hongmei Wang, Lin Chen, et al. "Safety of Fetal Olfactory Ensheathing Cell Transplantation in Patients with Chronic Spinal Cord Injury: A 38-Month Follow-Up with MRI." *Zhongguo Xiu Fu Chong Jian Wai Ke Za Zhi* 20, no. 4 (2006): 439–43.

Huang, Xi-Ping, Zhuo Sun, Yasuo Miyagi, et al. "Differentiation of Allogeneic Mesenchymal Stem Cells Induces Immunogenicity and Limits Their Long-Term Benefits for Myocardial Repair." *Circulation* 122 (2010): 2419–29.

Huang, Xiaosong, Scott Cho, and Gerald J. Spangrude. "Hematopoietic Stem Cells: Generation and Self-Renewal." *Cell Death and Differentiation* 14, no. 11 (2007): 1851–9.

Huard, Josee M.T., Steven L. Youngentob, Bradley J. Goldstein, Marla B. Luskin, and James E. Schwob. "Adult Olfactory Epithelium Contains Multipotent Progenitors that Give Rise to Neurons and Non-Neuronal Cells." *Comparative Neurology* 400 (1998): 469–86.

Hui, Edwin. *At the Beginning of Life*. Downers Grove, IL: Intervarsity Press, 2002.

Humanity+. http://humanityplus.org.

Humpherys, David, Kevin Eggan, Hidenori Akutsu, et al. "Abnormal Gene Expression in Cloned Mice Derived from Embryonic Stem Cell and Cumulus Cell Nuclei." *Proceedings of the National Academy of Sciences USA* 99, no. 20 (2002): 12889–94.

Hurlbut, William B. "Altered Nuclear Transfer as a Morally Acceptable Means for the Procurement of Human Embryonic Stem Cells." *Perspectives in Biology and Medicine*, 48, no. 2 (2005): 211–28.

Hurlbut, William B., Robert P. George, and Markus Gompe. "Seeking Consensus: A Clarification and Defense of Altered Nuclear Transfer." *Hastings Center Report* 36, no. 5 (2006): 42–50.

Huser, Roger J. *The Crime of Abortion in Canon Law*. Washington, D.C.: The Catholic University of America Press, 1942.

Hussein, Samer, Nizar N. Batada, Sanna Vuoristo, et al. "Copy Number Variation and Selection During Reprogramming to Pluripotency." *Nature* 471 (2011): 58–62.

Hutchinson, Martin. "Aborted Foetus Could Provide Eggs?" *BBC News*. July 16, 2003. http://news.bbc.co.uk/2/hi/health/3031800.stm.

Hüttmann, Andreas, Chung-Leung Li, and Ulrich Dührsen. "Bone Marrow-Derived Stem Cells and'Plasticity,'" *Annals of Hematology* 82, no. 10 (2003): 599–604.

Hwang, William,Ying Khee, Miny Samuel, Daryl Tan, Liang Piu Koh, Winston Lim, and Yeh Ching Linn "A Meta-Analysis of Unrelated Donor Umbilical Cord Blood Transplantation Versus Unrelated Donor Bone Marrow Transplantation in Adult and Pediatric Patients." *Biology of Blood and Bone Marrow Transplantation* 13, no. 4 (2007): 444–53.

Ichim, Thomas E., F. Solano, R. Brenes, et al. "Placental Mesenchymal and Cord Blood Stem Cell Therapy for Dilated Cardiomyopathy." *Reproductive Biomedicine Online* 16, no. 6 (2008): 898–905.

Ichim, Thomas E., Fabio Solano, Fabian Lara, et al. "Feasibility of Combination Allogeneic Stem Cell Therapy of Spinal Cord Injury: A Case Report." *International Archives of Medicine* 3 (2010): 30.

Igura, K., X. Zhang, K. Takahashi, A. Mitsuru, S. Yamaguchi, and T. A. Takahashi. "Isolation and Characterization of Mesenchymal Progenitor Cells from Chorionic Villi of Human Placenta." *Cytotherapy* 6 (2004): 543–53.

Ilancheran, Sivakami, Anna Michalska, Gary Peh, Euan M. Wallace, Martin Pera, and Ursula Manuelpillai. "Stem Cells Derived from Human Fetal Membranes Display Multilineage Differentiation Potential." *Biology of Reproduction* 77 (2007): 577–88.

in 't Anker, Pieternella S., Willy A. Noort, Sicco A. Scherjon, et al. "Mesenchymal Stem Cells in Human Second-Trimester Bone Marrow, Liver, Lung, and Spleen Exhibit a Similar Immunophenotype but a Heterogeneous Multilineage Differentiation Potential." *Haematologica* 88 (2003): 845–52.

Institute of Medicine. *Assessing the Medical Risks of Human Oocyte Donation for Stem Cell Research: Workshop Report*. Washington DC: National Academic Press, 2007.

Inui, Akio. "Obesity—A Chronic Health Problem in Cloned Mice?" *Trends in Pharmacological Science* 24 (2003): 77–80.

Isaev, Dmitry A., Ibon Garitaonandia, Tatiana V. Abramihina, Tatjana Zogovic-Kapsalis, Richard A. West, Andrey Y. Semechkin, Albrecht M. Müller, and Ruslan A. Semechkin. "*In Vitro* Differentiation of Human Parthenogenetic Stem Cells into Neural Lineages." *Regenerative Medicine* 7 (2012): 37–45.

Itzhaki, Ilanit, Sophia Rapoport, Irit Huber, et al. "Calcium Handling in Human Induced Pluripotent Stem Cell Derived Cardiomyocytes." *PLoS One* 6, no. 4 (2011): e18037. doi:10.1371/journal.pone.0018037

Jaenisch, Rudolf, "The Biology of Nuclear Cloning and the Potential of Embryonic Stem Cells for Transplantation Therapy." In *President's Council of Bioethics, Monitoring Stem Cell Research*, 387–417. Washington, D.C.: President's Council on Bioethics, 2004. Available at http://bioethics.georgetown.edu/pcbe/reports/stemcell/appendix_n.html.

Jain, Mohit, Harout DerSimonian, Daniel A. Brenner, et al. "Cell Therapy Attenuates Deleterious Ventricular Remodeling and Improves Cardiac Performance after Myocardial Infarction." *Circulation* 103, no. 14 (2001): 1920–7.

James, Sarah. "The Painful Truth." Interview with Maria Shriver. *Dateline*. NBC, January 2, 2000.

Jeong, Soo Ryeong, Min Jung Kwon, Hwan Goo Lee, et al. "Hepatocyte Growth Factor Reduces Astrocytic Scar Formation and Promotes Axonal Growth Beyond Glial Scars after Spinal Cord Injury." *Experimental Neurology* 233 (2012): 312–22.

Jessup, Mariell and Susan Brozena. "Guidelines for the Management of Heart Failure: Differences and Guidelines Perspectives." *Cardiology Clinics* 25, no. 4 (2007): 497–506.

Jiang, Meng, Ben He, Qi Zhan, et al. "Randomized Controlled Trials on the Therapeutic Effects of Adult Progenitor Cells for Myocardial Infarction: Meta-Analysis." *Expert Opinion on Biological Therapy* 10, no. 5 (2010): 667–80.

Jiang, Yuehua, Balkrishna N. Jahagirdar, R. Lee Reinhardt, et al. "Pluripotency of Mesenchymal Stem Cells Derived from Adult Marrow." *Nature* 418 (2002): 41–9.

Johnson, Leonard F. "Placental Blood Transplantation and Autologous Banking–Caveat Emptor." *Journal of Pediatric Hematology/Oncology* 19, no. 3 (1997): 183–6.

Johnston, Peter V., Tetsuo Sasano, Kevin Mills, et al. "Engraftment, Differentiation and Functional Benefits of Autologous Cardiosphere-Derived Cells in Porcine Ischemic Cardiomyopathy." *Circulation* 120 (2009): 1075–83.

Jones, David Albert. *The Soul of the Embryo: An Enquiry into the Status of the Human Embryo in the Christian Tradition*. New York: Continuum, 2004.

Jones, Gemma N., Dafni Moschidou, Tamara-Isabel Puga-Iglesias, et al. "Ontological Differences in First Compared to Third Trimester Human Fetal Placental Chorionic Stem Cells." *PLoS One* 7, no. 2 (2012): e43395.

Jones, Howard W. "Multiple Births: How Are We Doing?" *Fertility and Sterility* 79, no. 1 (2003): 17–21.

Joraku, Akira, Kathryn A. Stern, Anthony Atala, and James J. Yoo. "*In vitro* Generation of Three-Dimensional Renal Structures." *Methods* 47, no. 2 (2009): 129–33.

Joyce, Nancy C., Deshea L. Harris, Vladimir Markov, Zhe Zhang, and Biagio Saitta. "Potential of Human Umbilical Cord Blood Mesenchymal Stem Cells to Heal Damaged Corneal Endothelium." *Molecular Vision* 18 (2012): 547–64.

Jujo, Kentaro, Masaaki Ii, and Douglas W. Losordo. "Endothelial Progenitor Cells in Neovascularization of Infarcted Myocardium." *Journal of Molecular and Cellular Cardiology* 45, no. 4 (2008): 530–44.

Kajstura, Jan, Marcello Rota, Brian Whang, et al. "Bone Marrow Cells Differentiate in Cardiac Cell Lineages after Infarction Independently of Cell Fusion." *Circulation Research* 96 (2005): 127–37.

Kalman, Matthew. "A Stem-Cell Prospect for Ailing Hearts." *Time Magazine Online*, Nov 14, 2007. http://www.time.com/time/magazine/article/0,9171,1137670,00.html.

Kang, K.-S., S. W. Kim, Y. H. Oh, et al. "A 37-Year-Old Spinal Cord-Injured Female Patient, Transplanted of Multipotent Stem Cells from Human UC Blood, with Improved Sensory Perception and Mobility, Both Functionally and Morphologically: A Case Study." *Cytotherapy* 7, no. 4 (2005): 368–73.

Kato, Yoko, Tetsuya Tani, and Yukio Tsunoda. "Cloning of Calves from Various Somatic Cell Types of Male and Female Adult, Newborn and Fetal Cows." *Journal of Reproduction and Fertility* 120 (2000): 231–7.

Katritsis, Demosthenes G., Panagiota A. Sotiropoulou, Evangelia Karvouni, et al. "Transcoronary Transplantation of Autologous Mesenchymal Stem Cells and Endothelial Progenitors into Infarcted Human Myocardium." *Catheterization and Cardiovascular Interventions* 65, no. 3 (2005): 321–9.

Kawamoto, Atsuhiko, Tengis Tkebuchava, Jun-Ichi Yamaguchi, et al. "Intramyocardial Transplantation of Autologous Endothelial Progenitor Cells for Therapeutic Neovascularization of Myocardial Ischemia." *Circulation* 107, no. 3 (2003): 461–8.

Kay, Katty, and Mark Henderson. "Paralyzed Mouse Walks Again as Scientists Fight Stem Cell Ban." In *The Stem Cell Controversy*, 2nd edition, edited by Michael Ruse and Christopher A. Pynes, 73–5. Amherst, NY: Prometheus Press, 2006.

Keefer, Carol L., R. Keyston, A. Lazaris, et al. "Production of Cloned Goats after Nuclear Transfer Using Adult Somatic Cells." *Biology of Reproduction* 66 (2002): 199–203.

Kehat, Izhak, Dorit Kenyagin-Karsenti, Mirit Snir, et al. "Human Embryonic Stem Cells Can Differentiate into Myocytes with Structural and Functional Properties of Cardiomyocytes." *Journal of Clinical Investigation* 108, no. 3 (2001): 407–14.

Keirstead, Hans S., and Gabriel Nistor, Giovanna Bernal, et al. "Human Embryonic Stem Cell-Derived Oligodendrocytes Progenitor Cell Transplants Remyelinate and Restore Locomotion after Spinal Cord Injury." *Journal of Neuroscience* 25 (2005): 4694–705.

Kern, Susanne, Hermann Eichler, Johannes Stoeve, Harald Klüter, and Karen Bieback. "Comparative Analysis of Mesenchymal Stem Cells from Bone Marrow, Umbilical Cord Blood, or Adipose Tissue." *Stem Cells* 24 (2006): 1294–1301.

Kerr, Douglas A., Jerònia Lladó, Michael J. Shamblott, et al. "Human Embryonic Germ Cell Derivatives Facilitate Motor Recovery of Rats with Diffuse Motor Neuron Injury." *Journal of Neuroscience* 23 (2003): 5131–40.

Keyvan-Fouladi, Naghmeh, Geoffrey Raisman, and Ying Li. "Functional Repair of the Corticospinal Tract by Delayed Transplantation of Olfactory Ensheathing Cells in Adult Rats." *Journal of Neuroscience* 23, no. 28 (2003): 9428–34.

Khouri, Issa F. "Allogeneic Stem Cell Transplantation in Follicular Lymphoma." *Best Practice and Research. Clinical Haematology* 24, no. 2 (2011): 271–7.

Kiessling, Ann A., and Scott C. Anderson. *Human Embryonic Stem Cells.* 2nd ed. Sudbury, MA: Jones and Bartlett, 2007.

Kilner, John F., and C. Ben Mitchell. *Does God Need Our Help? Cloning, Assisted Suicide, & Other Challenges in Bioethics.* Wheaton, IL: Tyndale House Publishers, 2003.

Kim, Byung Gon, Young Mi Kang, Ji Hoon Phi, et al. "Implantation of Polymer Scaffolds Seeded with Neural Stem Cells in a Canine Spinal Cord Injury Model." *Cytotherapy* 12, no. 6 (2010): 841–5.

Kim, Byung-Ok, Hai Tian, Kriengchai Prasongsukarn, et al. "Cell Transplantation Improves Ventricular Function after a Myocardial Infarction: A Preclinical Study of Human Unrestricted Somatic Stem Cells in a Porcine Model." supplement, *Circulation* 112, no. S9 (2005): I96–I104.

Kim, Kitai, Paul Lerou, Akiko Yabuuchi, et al. "Histocompatible Embryonic Stem Cells by Parthenogenesis." *Science* 315 (2007): 482–6.

Kirkegaard, Kirstine, Johnny Juhl Hindkjaer, and Hans Jacob Ingerslev. "Human Embryonic Development After Blastomere Removal: A Time-Lapse Analysis." *Human Reproduction* 27 (2012): 97–105.

Kishi, Masao, Y. Itagaki, R. Takakura, et al. "Nuclear Transfer in Cattle Using Colostrum-Derived Mammary Gland Epithelial Cells and Ear-Derived Fibroblast Cells." *Theriogenology* 54 (2000): 675–84.

Klassen, Henry, Boback Ziaeian, Ivan I. Kirov, Michael J. Young, and Philip H. Schwartz. "Isolation of Retinal Progenitor Cells from Post-Mortem Human Tissue and Comparison with Autologous Brain Progenitors." *Journal of Neuroscience Research* 77, no. 3 (2004): 334–43.

Klein, H. M., A. Ghodsizad, R. Marktanner, et al. "Intramyocardial Implantation of CD133+ Stem Cells Improved Cardiac Function without Bypass Surgery." *Heart Surgery Forum* 10, no. 1 (2007): E66–9.

Klimanskaya, Irina, Young Chung, Sandy Becker, Shi-Jiang Lu, and Robert Lanza. "Human Embryonic Stem Cell Lines Derived from Single Blastomeres." *Nature* 444 (2006): 481–5.

———. "Derivation of Human Embryonic Stem Cells from Single Blastomeres." *Nature Protocols* 2, no. 8 (2007): 1963–72.

Klock, Susan C., Sandra Sheinin, and Ralph R. Kazer. "The Disposition of Unused Frozen Embryos." *New England Journal of Medicine* 345 (2001): 69–70.

Klonoff-Cohen, Hillary S., and Loki Natarajan. "The Effect of Advancing Paternal Age on Pregnancy and Live Birth Rates in Couples Undergoing In Vitro Fertilization or Gamete Intrafallopian Transfer." *American Journal of Obstetrics and Gynecology* 191, no. 2 (2004): 507–14.

Klusendorf, Scott. *The Case for Life: Equipping Christians to Engage the Culture.* Wheaton, IL:Crossway Books, 2009.

Kobylka, Peter, Pavol Ivanyi, and Birgitta S. Breur-Vriesendorp. "Preservation of Immunological and Colony-Forming Capacities of Long-Term (15 Years) Cryo-preserved Cord Blood Cells." *Transplantation* 65, no. 9 (1998): 1275–8.

Koç, Omer N., Stanton L. Gerson, Brenda W. Cooper, et al. "Rapid Hematopoietic Recovery After Coinfusion of Autologous-Blood Stem Cells and Culture-Expanded Marrow Mesenchymal Stem Cells in Advanced Breast Cancer Patients Receiving High-Dose Chemotherapy." *Journal of Clinical Oncology* 18, no. 2 (2000): 307–16.

Kocher, A. A., M. D. Schuster, M. J. Szabolcs, et al. "Neovascularization of Ischemic Myocardium by Human Bone Marrow-Derived Angioblasts Prevents Cardiomyocyte Apoptosis, Reduces Remodeling and Improves Cardiac Function." *Nature Medicine* 7, no. 4 (2001): 430–6.

Kofidis, Theo, Jorg Lucas deBruin, Masashi Tanaka, et al. "They Are Not Stealthy in the Heart: Embryonic Stem Cells Trigger Cell Infiltration, Humoral and T-Lymphocyte-Based Host Immune Response." *European Journal of Cardio-Thoracic Surgery* 28, no. 3 (2005): 461–6.

Kögler, Gesine, Sandra Sensken, Judith A. Airey, et al. "A New Human Somatic Stem Cell from Placental Cord Blood with Intrinsic Pluripotent Differentiation Potential." *Journal of Experimental Medicine* 200, no. 2 (2004): 123–35.

Kohama, Ikuhide, Karen L. Lankford, Jana Preiningerova, Fletcher A. White, Timothy L. Vollmer, and Jeffery D. Kocsis. "Transplantation of Cryopreserved Adult Human

Schwann Cells Enhances Axonal Conduction in Demyelinated Spinal Cord." *Journal of Neuroscience* 21, no. 3 (2001): 944–50.

Kohda, Takashi, Kimiko Inoue, Narumi Ogonuki, et al. "Variation in Gene Expression and Aberrantly Regulated Chromosome Regions in Cloned Mice." *Biology of Reproduction* 73 (2005): 1302–11.

Kojima, Atsuhiro, and Charles H. Tator. "Intrathecal Administration of Epidermal Growth Factor and Fibroblast Growth Factor-2 Promotes Ependymal Proliferation and Functional Recovery after Spinal Cord Injury in Adult Rats." *Journal of Neurotrauma* 19 (2002): 223–38.

Kolambkar, Yash M., Alexandra Peister, Shay Soker, Anthony Atala, and Robert E. Guldberg. "Chondrogenic Differentiation of Amniotic Fluid-Derived Stem Cells." *Journal of Molecular Histology* 38, no. 5 (2007): 405–13.

Kopen, Gene C., Darwin J. Prockop, and Donald G. Phinney. "Marrow Stromal Cells Migrate throughout Forebrain and Cerebellum, and They Differentiate into Astrocytes after Injection into Neonatal Mouse Brains." *Proceedings of the National Academy of Sciences USA* 96 (1999): 10711–6.

Kordower, Jeffrey H., Yaping Chu, Robert A. Hauser, Thomas B. Freeman, and C. Warren Olanow. "Lewy Body–Like Pathology in Long-Term Embryonic Nigral Transplants in Parkinson's Disease." *Nature Medicine* 14 (2008): 504–6.

Koukl, Gregory. *Precious Unborn Human Persons*. Signal Hill, CA: Stand to Reason Press, 1999.

Kovacic, B., V. Vlaisavljevic, M. Reljic, and M. Cizek-Sajko. "Developmental Capacity of Different Morphological Types of Day 5 Human Morulae and Blastocysts." *Reproductive Biomedicine Online* 8, no. 6 (2004): 687–94.

Krauthammer, Charles. "An Edwards Outrage." *Washington Post*, Oct. 15, 2004.

Kreeft, Peter. *Three Approaches to Abortion: A Thoughtful and Compassionate Guide to Today's Most Controversial Issue*. San Francisco: Ignatius Press, 2002.

Kucia, Magda, Buddhadeb Dawn, Greg Hunt, et al. "Cells Expressing Early Cardiac Markers Reside in the Bone Marrow and are Mobilized into the Peripheral Blood after Myocardial Infarction." *Circulation Research* 95 (2004): 1191–9.

Laflamme, Michael A., Kent Y. Chen, Anna V. Naumova, et al. "Cardiomyocytes Derived from Human Embryonic Stem Cells in Pro-Survival Factors Enhance Function of Infarcted Rat Hearts." *Nature Biotechnology* 25 (2007): 1015–24.

Laflamme, Michael A., Joseph Gold, Chunhui Xu, et al. "Formation of Human Myocardium in the Rat Heart from Human Embryonic Stem Cells." *American Journal of Pathology* 167, no. 3 (2005): 663–71.

Lai, Liangxue, Donna Kolber-Simonds, Kwang-Wook Park, et al. "Production of α-1,3-Galactosyltransferase Knockout Pigs by Nuclear Transfer Cloning." *Science* 295 (2002): 1089–92.

Landry, Donald W., and Howard Zucker. "Embryonic Death and the Creation of Human Embryonic Stem Cells." *The Journal of Clinical Investigation* 114, no. 9 (2004): 1184–6.

Landry, Donald W., Howard A. Zucker, Mark V. Sauer, Michael Reznik, and Lauren Wiebe. "Hypocellularity and Absence of Compaction as Criteria for Embryonic Death." *Regenerative Medicine* 1, no. 3 (2006): 367–71.

Lanza, Cristina, Sara Morando, Adriana Voci, et al. "Neuroprotective Mesenchymal Stem Cells are Endowed with a Potent Antioxidant Effect In Vivo." *Journal of Neurochemistry* 110 (2009): 1674–84.

Lanza, Robert P., Jose B. Cibelli, David Faber, et al. "Cloned Cattle Can Be Healthy and Normal." *Science* 294 (2001): 1893–4.

Lanza, Robert P., Ho Yun Chung, James J. Yoo et al. "Generation of Histocompatible Tissues Using Nuclear Transplantation." *Nature Biotechnology* 20 (2002): 689–96.

Lanza, Robert, J.-H. Shieh, Peter J. Wettstein, et al. "Long-Term Bovine Hematopoietic Engraftment with Clone-Derived Stem Cells." *Cloning and Stem Cells* 7, no. 2 (2005): 95–106.

Lanza, Robert, Malcolm A.S. Moore, Teruhiko Wakayama, et al. "Regeneration of the Infarcted Heart with Stem Cells Derived by Nuclear Transplantation." *Circulation Research* 94 (2004): 820–7.

Laurent, Louise C., Igor Ulitsky, Ileana Slavin, et al. "Dynamic Changes in the Copy Number of Pluripotency and Cell Proliferation Genes in Human ESCs and iPSCs during Reprogramming and Time in Culture." *Cell Stem Cell* 8, no. 1 (2011): 106–18.

Laverge, H., J. Van der Elst, P. De Sutter, M.R. Verschraegen-Spae, A. De Paepe, and M. Dhont. "Fluorescent In-situ Hybridization on Human Embryos Showing Cleavage Arrest after Freezing and Thawing." *Human Reproduction* 13 (1998): 425–9.

Lazarus, Hillard M., S. E. Haynesworth, S. L. Gerson, N. S. Rosenthal, and A. I. Caplan. "Ex Vivo Expansion and Subsequent Infusion of Human Bone Marrow-Derived Stromal Progenitor Cells (Mesenchymal Progenitor Cells): Implications for Therapeutic Use." *Bone Marrow Transplantation* 16 (1995): 557–64.

Lazarus, Hillard M., Omer N. Koç, Steven M. Devine, et al. "Cotransplantation of HLA-Identical Sibling Culture-Expanded Mesenchymal Stem Cells and Hematopoietic Stem Cells in Hematologic Malignancy Patients." *Biology of Blood and Marrow Transplant* 11 (2005): 389–98.

Le Blanc, Katarina, and Mark Pittenger. "Mesenchymal Stem Cells: Progress Toward Promise." *Cytotherapy* 7 (2005): 36–45.

Lee, Byeong Chun, Min Kyu Kim, Goo Jang, et al. "Dogs Cloned from Adult Somatic Cells." *Nature* 436 (2005): 641.

Lee, Jiyoon, Andrea Wecker, Douglas W. Losordo, and Young-sup Yoon. "Derivation and Characterization of Bone Marrow-Derived Multipotent Stem Cells." *Experimental Hematology* 34, no. 11 (2006): 1602–3.

Lee, Patrick. *Abortion and Unborn Human Life*. Washington D.C.: Catholic University of America Press, 1997.

Lee, Seung-Tae, J. H. Jang, J. W. Cheong, et al. "Treatment of High-Risk Acute Myelogenous Leukaemia by Myeloablative Chemoradiotherapy Followed by Co-Infusion of T Cell-Depleted Haematopoietic Stem Cells and Culture-Expanded Marrow Mesenchymal Stem Cells from a Related Donor with One Fully Mismatched Human Leucocyte Antigen Haplotype." *British Journal of Haematology* 118 (2002): 1128–31.

Lee, Wen-Yu, Hao-Ji Wei, Wei-Wei Lin, et al. "Enhancement of Cell Retention and Functional Benefits in Myocardial Infarction Using Human Amniotic-Fluid Stem-Cell Bodies Enriched with Endogenous ECM." *Biomaterials* 32, no. 24 (2011): 5558–67.

Lengerke, Claudia, Kitai Kim, Paul Lerou, and George Q. Daley. "Differentiation Potential of Histocompatible Parthenogenetic Embryonic Stem Cells." *Annals of the New York Academy of Sciences* 1106 (2007): 209–18.

Léobon, Bertrand, Isabelle Garcin, Philippe Menasché, Jean-Thomas Vilquin, Etienne Audinat, and Serge Charpak. "Myoblasts Transplanted in Rat Infarcted Myocardium are Functionally Isolated from Their Host." *Proceedings of the National Academy of Sciences USA* 100, no. 13 (2003): 7808–11.

Leri, Annarosa, Jan Kajstura, and Piero Anversa. "Role of Cardiac Stem Cells in Cardiac Physiology: A Paradigm Shift in Human Myocardial Biology." *Circulation Research* 109 (2011): 941–61.

Lerou, Paul H., Akiko Yabuuchi, Hongguang Huo, et al. "Human Embryonic Stem Cell Derivation from Poor-Quality Embryos." *Nature Biotechnology* 26, no. 2 (2008): 212–4.

Lessenberry, Jack. "Stem Cell Lies." *Metro Times*, October 8, 2008. http://www2.metrotimes.com/editorial/story.asp?id=13325.

Lesný, Petr, M. Přádný, P. Jendelová, J. Michálek, J. Vacík and E. Syková. "Macroporous Hydrogels Based on 2-Hydroxyethyl Methacrylate. Part 4: Growth of Rat Bone Marrow Stromal Cells in Three-Dimensional Hydrogels with Positive and Negative Surface Charges and in Polyelectrolyte Complexes." *Journal of Material Science: Materials in Medicine* 17, no. 9 (2006): 829–33.

Lester, Natalie. "Embryo Adoption Becoming the Rage." *The Washington Times*, April 19, 2009. http://www.washingtontimes.com/news/2009/apr/19/embryo-adoption-becoming-rage.

Li, Chang Dong, Wei Yuan Zhang, He Lian Li, Xiao Xia Jiang, Yi Zhang, Pei Hsien Tang, and Ning Mao. "Mesenchymal Stem Cells Derived from Human Placenta Suppress Allogeneic Umbilical Cord Blood Lymphocyte Proliferation." *Cell Research* 15 (2005): 539–47.

Li, Haochuan, Jerry Y. Niederkorn, Sudha Neelam, Elizabeth Mayhew, R. Ann Word, James P. McCulley, and Hassan Alizadeh. "Immunosuppressive Factors Secreted by Human Amniotic Epithelial Cells." *Investigative Ophthalmology and Visual Science* 46, no. 3 (2005): 900–7.

Li, Jia-Yi, Elisabet Englund, Janice L Holton, et al. "Lewy Bodies in Grafted Neurons in Subjects with Parkinson's Disease Suggest Host-to-Graft Disease Propagation." *Nature Medicine* 14 (2008): 501–3.

Li, Li, Miren L. Baroja, Anish Majumdar, et al. "Human Embryonic Stem Cells Possess Immuno-Privileged Properties." *Stem Cells* 22 (2004): 448–56.

Li, Wenzhong, Nan Ma, Lee-Lee Ong, et al. "Bcl-2 Engineered MSCs Inhibited Apoptosis and Improved Heart Function." *Stem Cells* 25, no. 8 (2007): 2118–27.

Li, Yan, Shi Chen, Jin Yuan, et al. "Mesenchymal Stem/Progenitor Cells Promote the Reconstitution of Exogenous Hematopoietic Stem Cells in *Fancg-/-* Mice in Vivo." *Blood* 113, no. 10 (2009): 2342–51.

Li, Ying, Patrick Decherchi, and Geoffrey Raisman. "Transplantation of Olfactory Ensheathing Cells into Spinal Cord Lesions Restores Breathing and Climbing." *Journal of Neuroscience* 23, no. 3 (2003): 727–31.

Liao, Song-Yan, Yuan Liu, Chung-Wah Siu, et al. "Proarrhythmic Risk of Embryonic Stem Cell-Derived Cardiomyocyte Transplantation in Infarcted Myocardium." *Heart Rhythm* 7, no. 12 (2010): 1853–9.

Liao, Yanling, Mark B. Geyer, Albert J. Yang, and Mitchell S. Cairo. "Cord Blood Transplantation and Stem Cell Potential." *Experimental Hematology* 39 (2011): 393–412.

Liechty, Kenneth W., Tippi C. MacKenzie, Aimen F. Shaaban, et al. "Human Mesenchyme Stem Cells Engraft and Demonstrate Site-Specific Differentiation after *In Utero* Transplantation in Sheep." *Nature Medicine* 6, no. 11 (2000): 1282–6.

Lightfoot, J. B. *The Apostolic Fathers*. Edited by J. R. Harmer. Translated by J. B. Lightfoot. Grand Rapids, MI: Baker Book House, 1973.

Lim, Sang Yup, Yong Sook Kim, Youngkeun Ahn, et al. "The Effects of Mesenchymal Stem Cells Transduced with AKT in a Porcine Myocardial Infarction Model." *Cardiovascular Research* 70 (2006): 530–42.

Lima, Carlos, José Pratas-Vital, Pedro Escada, Armando Hasse-Ferreira, Clara Capucho, and Jean D Peduzzi. "Olfactory Mucosa Autografts in Human Spinal Cord Injury: A Pilot Clinical Study." *Journal of Spinal Cord Medicine* 29 (2006): 191–203.

Lin, Qiuxia, Qiang Fu, Ye Zhang. "Tumorigenesis in the Infarcted Rat Heart is Eliminated through Differentiation and Enrichment of the Transplanted Embryonic Stem Cells." *European Journal of Heart Failure* 12, no. 11 (2010): 1179–85.

Linke, Axel, Patrick Müller, Daria Nurzynska, et al. "Stem Cells in the Dog Heart Are Self-Renewing, Clonogenic, and Multipotent and Regenerate Infarcted Myocardium, Improving Cardiac Function." *Proceedings of the National Academy of Sciences USA* 102, no. 25 (2005): 8966–71.

Lister, Ryan, Mattia Pelizzola, Yasuyuki S. Kida, et al. "Hotspots of Aberrant Epigenomic Reprogramming in Human Induced Pluripotent Stem Cells." *Nature* 471 (2011): 68–73.

Liu, Hongshan, Jianhua Zhang, Chia-Yang Liu, et al. "Cell Therapy of Congenital Corneal Diseases with Umbilical Mesenchymal Stem Cells: Lumican Null Mice." *PLoS ONE* 5, no. 5 (2010): e10707. doi:10.1371/journal.pone.0010707.

Lodi, Daniele, Tommaso Iannitti, and Beniamino Palmieri. "Stem Cells in Clinical Practice: Applications and Warnings." *Journal of Experimental and Clinical Cancer Research* 30, no. 1 (2011): 9.

Loffredo, Francesco S., Matthew L. Steinhauser, Joseph Gannon, and Richard T. Lee. "Bone Marrow-Derived Cell Therapy Stimulates Endogenous Cardiomyocyte Progenitors and Promotes Cardiac Repair." *Cell Stem Cell* 8 (2011): 389–98.

Loi, Pasqualino, Grazyna Ptak, Barbara Barboni, Josef Fulka, Pietro Cappai, and Michael Clinton. "Genetic Rescue of an Endangered Mammal by Cross-Species Nuclear Transfer Using Post-Mortem Somatic Cells." *Nature Biotechnology* 19 (2001): 962–4.

Long, Jian-Er, Xia Cai, and Li-Qiang He. "Gene Profiling of Cattle Blastocysts Derived from Nuclear Transfer, *In Vitro* Fertilization and In Vivo Development Based on cDNA Library." *Animal Reproductive Science* 100, no. 3-4 (2007): 243–56.

Longhi, Luca, Elisa R. Zanier, Nicolas Royo, Nino Stocchetti, and Tracy K. McIntosh. "Stem Cell Transplantation as a Therapeutic Strategy for Traumatic Brain Injury." *Transplantation Immunology* 15 (2005): 143–8.

Lorthongpanich, Chanchao, Shang-Hsun Yang, Karolina Piotrowska-Nitsche, Rangsun Parnpai, and Anthony W. S. Chan. "Development of Single Mouse Blastomeres into Blastocysts, Outgrowths and the Establishment of Embryonic Stem Cells." *Reproduction* 135 (2008): 805–13.

Losordo, Douglas W., Timothy D. Henry, Charles Davidson, et al. "Intramyocardial, Autologous CD34+ Cell Therapy for Refractory Angina." *Circulation Research* 109, no. 4 (2011): 428–36.

Lu, Hui-he, Yi-fei Li, Zheng-qiang Sheng, and Yi Wang. "Preconditioning of Stem Cells for the Treatment of Myocardial Infarction." *Chinese Medical Journal* 125, no. 2 (2012): 378–84.

Lu, Jike, François Féron, Alan Mackay-Sim, and Phil M. E. Waite. "Olfactory Ensheathing Cells Promote Locomotor Recovery after Delayed Transplantation into Transected Spinal Cord." *Brain* 125 (2002): 14–21.

Lund, Raymond D., Shaomei Wang, Bin Lu, et al. "Cells Isolated from Umbilical Cord Tissue Rescue Photoreceptors and Visual Functions in a Rodent Model of Retinal Disease." *Stem Cells* 25 (2007): 602–11.

Lundin, Susanne. "'I Want a Baby; Don't Stop Me from Being a Mother': An Ethnographic Study on Fertility Tourism and Egg Trade." *Cultural Politics* 8, no. 2 (2012): 327–44.

Lyerly, Anne Drapkin, and Rith R. Faden. "Willingness to Donate Frozen Embryos for Stem Cell Research." *Science* 317 (2007): 46–7.

Lyerly, Anne Drapkin, Karen Steinhauser, Corrine Voils, et al. "Fertility Patients' Views About Frozen Embryo Disposition: Results of a Multi-Institutional U.S. Survey." *Fertility and Sterility* 93, no. 2 (2010): 499–509.

Macchiarini, Paolo, Philipp Jungebluth, Tetsuhiko Go, et al. "Clinical Transplantation of a Tissue–Engineered Airway." *Lancet* 372 (2008): 2023–30.

Machin, Geoffrey. "Familial Monozygotic Twinning: A Report of Seven Pedigrees." *American Journal of Medical Genetics Part C: Seminars in Medical Genetics* 151C, no. 2 (2009): 152–4.

Mackay-Sim, Alan, F. Féron, J. Cochrane, et al. "Autologous Olfactory Ensheathing Cell Transplantation in Human Paraplegia: A 3-Year Clinical Trial." *Brain* 131 (2008): 2376–86.

Magureanu, George. "Human Egg Trading and the Exploitation of Women." Presentation at the CORE European Seminar, European Parliament, Brussels. June 30, 2005. http://www.handsoffourovaries.com/pdfs/appendixg.pdf.

Maher, Brendan. "Egg Shortage Hits Race to Clone Human Stem Cells." *Nature* 453 (2008): 828–9.

Majumdar, Manas K., M. Keane-Moore, D. Buyaner, et al. "Characterization and Functionality of Cell Surface Molecules on Human Mesenchymal Stem Cells." *Journal of Biomedical Science* 10, no. 2 (2003): 228–41.

Makino, Shinji, Keiichi Fukuda, Shunichirou Miyoshi, et al. "Cardiomyocytes Can Be Generated from Marrow Stromal Cells *In Vitro*." *Journal of Clinical Investigation* 103, no. 6 (1999): 697–705.

Makkar, Raj R., Matthew J. Price, Michael Lill, et al. "Intramyocardial Injection of Allogenic Bone Marrow-Derived Mesenchymal Stem Cells without Immunosuppression Preserves Cardiac Function in a Porcine Model of Myocardial Infarction." *Journal of Cardiovascular Pharmacology and Therapeutics* 10, no. 4 (2005): 225–33.

Makkar, Raj R., Rachel R. Smith, Ke Cheng, et al. "Intracoronary Cardiosphere-Derived Cells for Heart Regeneration after Myocardial Infarction (CADUCEUS): A Prospective, Randomised Phase 1 Trial." *Lancet* 379 (2012): 895–904.

Mali, Prashant, Bin-Kuan Chou, Jonathan Yen, et al. "Butyrate Greatly Enhances Derivation of Human Induced Pluripotent Stem Cells by Promoting Epigenetic Remodeling and the Expression of Pluripotency-Associated Genes." *Stem Cells* 28, no. 4 (2010): 713–20.

Mann, Mellissa R.W., Young Gie Chung, Leisha D. Nolen, Raluca I. Verona, Keith E. Latham, and Marisa S. Bartolomei. "Disruption of Imprinted Gene Methylation and Expression in Cloned Preimplantation Stage Mouse Embryos." *Biology of Reproduction* 69, no. 3 (2003): 902–14.

Mansour, Samer, Denis-Claude Roy, Vincent Bouchard, et al. "COMPARE-AMI Trial: Comparison of Intracoronary Injection of CD133+ Bone Marrow Stem Cells to Placebo in Patients After Acute Myocardial Infarction and Left Ventricular Dysfunction: Study Rationale and Design." *Journal of Cardiovascular Translational Research* 3, no. 2 (2010): 153–9.

Marcus, Akiva J., and Dale Woodbury. "Fetal Stem Cells from Extra-Embryonic Tissues: Do Not Discard." *Journal of Cellular and Molecular Medicine* 12, no. 3 (2008): 730–742.

Marion, Rosa M., Katerina Strati, Han Li, et al. "Telomeres Acquire Embryonic Stem Cell Characteristics in Induced Pluripotent Stem Cells." *Cell Stem Cell* 4, no. 2 (2009): 141–54.

Márquez, Carmen, Mireia Sandalinas, Muhterem Bahç, Mina Alikani, and Santiago Munné. "Chromosome Abnormalities in 1255 Cleavage-Stage Embryos." *Reproductive BioMedicine Online* 1 (2000): 17–27.

Martens, David J., Raewyn M. Seaberg, and Derek van der Kooy. "In Vivo Infusions of Exogenous Growth Factors in to the Fourth Ventricle of the Adult Mouse Brain Increase the Proliferation of Neural Precursors Around the Fourth Ventricle and the Central Canal of the Spinal Cord." *European Journal of Neuroscience* 16 (2002): 1045–57.

Martin, Gail R. "Isolation of a Pluripotent Cell Line from Early Mouse Embryos Cultured in Medium Conditioned by Teratocarcinoma Stem Cells." *Proceedings of the National Academy of Sciences USA* 78, no. 12 (1981): 7634–8.

Martins-Taylor, Kristen, Benjamin S. Nisler, Seth M. Taapken, et al. "Recurrent Copy Number Variations in Human Induced Pluripotent Stem Cells." *Nature Biotechnology* 29 (2011): 488–91.

Matsuse, Dai, Masaaki Kitada, Misaki Kohama, et al. "Human Umbilical Cord-Derived Mesenchymal Stromal Cells Differentiate Into Functional Schwann Cells That Sustain Peripheral Nerve Regeneration." *Journal of Neuropathology & Experimental Neurology* 69, no. 9 (2010): 973–85.

McCreath, Kenneth J., J. Howcroft, K. H. S. Campbell, A. Colman, A. E. Schnieke, and A. J. Kind. "Production of Gene-Targeted Sheep by Nuclear Transfer from Cultured Somatic Cells." *Nature* 405 (2000): 1066–9.

McCullough, Jeffrey, and Mary Clay. "Reasons for Deferral of Potential Umbilical Cord Blood Donors." *Transfusion* 40, no. 1 (2000): 124–5.

McHugh, Paul R. "Zygote and 'Clonote' – The Ethical Use of Embryonic Stem Cells." *New England Journal of Medicine* 351 (2004): 209–11.

McIntosh, Kevin R., and Amelia Bartholomew. "Stromal Cell Modulation of the Immune System: A Potential Role for Mesenchymal Stem Cells." *Graft* 3 (2000): 324–8.

McMahon, C., F. Gibson, J. Cohen, G. Leslie, C. Tennant, and D. Saunders. "Mothers Conceiving Through In Vitro Fertilization: Siblings, Setbacks, and Embryo Dilemmas After Five Years." *Reproductive Technologies* 10 (2000): 131–5.

McMahon, Catherine A., Frances L. Gibson, Garth I. Leslie, Douglas M. Saunders, Katherine A. Porter, and Christopher C. Tennant. "Embryo Donation for Medical Research: Attitude and Concerns of Potential Donors." *Human Reproduction* 18, no. 4 (2003): 871–7.

McNiece, Ian. "Delivering Cellular Therapies: Lessons Learned from *Ex Vivo* Culture and Clinical Applications of Hematopoietic Cells." *Seminars in Cell and Developmental Biology* 18, no. 6 (2007): 839–845.

Meissner, Alexander, and Rudolf Jaenisch. "Mammalian Nuclear Transfer." *Developmental Dynamics* 235 (2006): 2460–9.

———. "Generation of Nuclear Transfer-Derived Pluripotent ES Cell from Cloned Cdx-2-Deficient Blastocysts." *Nature* 439 (2006): 212–15.

Melton, Douglas A., George Q. Daley, and Charles G. Jennings. "Altered Nuclear Transfer in Stem-Cell Research—A Flawed Proposal." *New England Journal of Medicine* 351 (2004): 2791–2.

Ménard, Claudine, Albert A. Hagège, Onnik Agbulut, et al. "Transplantation of Cardiac-Committed Mouse Embryonic Stem Cells to Infarcted Sheep Myocardium: A Preliminary Study." *Lancet* 366 (2005): 1005–12.

Menasché, Philippe, Albert A. Hagège, Jean-Thomas Vilquin, et al. "Autologous Skeletal Myoblast Transplantation for Severe Postinfarction Left Ventricular Dysfunction." *Journal of the American College of Cardiology* 41 (2003): 1078–83.

Menasché, Philippe, Ottavio Alfieri, Stefan Janssens, et al. "The Myoblast Autologous Grafting in Ischemic Cardiomyopathy (MAGIC) Trial: First Randomized Placebo-Controlled Study of Myoblast Transplantation." *Circulation* 117, no. 9 (2008): 1189–1200.

Mendez, Ivar, Angel Viñuela, Arnar Astradsson, et al. "Dopamine Neurons Implanted into People with Parkinson's Disease Survive without Pathology for 14 Years." *Nature Medicine* 14 (2008): 507–9.

Messina, Elisa, Luciana De Angelis, Giacomo Frati, et al. "Isolation and Expansion of Adult Cardiac Stem Cells from Human and Murine Heart." *Circulation Research* 95 (2004): 911–21.

Meuleman, N., G. Vanhaelen, T. Trondreau, et al. "Reduced Intensity Conditioning Haematopoietic Stem Cell Transplantation with Mesenchymal Stromal Cells Infusion for the Treatment of Metachromatic Leukodystrophy: A Case Report." *Haematologica* 93 (2008): e11–13.

Mezey, É., B. Mayer, and K. Németh. "Unexpected Roles for Bone Marrow Stromal Cells (or MSCs): A Real Promise for Cellular, But Not Replacement Therapy." *Oral Diseases* 16, no. 2 (2010): 129–35.

Mezey, Éva, Sharon Key, Georgia Vogelsang, Ildiko Szalayova, G. David Lange, and Barbara Crain. "Transplanted Bone Marrow Generates New Neurons in Human Brains." *Proceedings of the National Academy of Sciences USA* 100, no. 3 (2003): 1364–9.

Michaelmann, H. W., and P. Nayudu. "Cryopreservation of Human Embryos." *Cell and Tissue Banking* 7, no. 2 (2006): 135–41.

Min, Jiang-Yong, Yinke Yang, Kimber L. Converso, et al. "Transplantation of Embryonic Stem Cells Improves Cardiac Function in Postinfarcted Rats." *Journal of Applied Physiology* 92 (2002): 288–96.

Mitsui, Kaoru, Yoshimi Tokuzawa, Hiroaki Itoh, et al. "The Homeoprotein Nanog Is Required for Maintenance of Pluripotency in Mouse Epiblast and ES Cells." *Cell* 113, no. 5 (2003): 631–42.

Miura, Kyoko, Yohei Okada, Takashi Aoi, et al. "Variation in the Safety of Induced Pluripotent Stem Cell Lines." *Nature Biotechnology* 27 (2009): 743–5.

Miura, Masako, Stan Gronthos, Mingrui Zhao, et al. "SHED: Stem Cells from Human Exfoliated Deciduous Teeth." *Proceedings of the National Academy of Sciences USA* 100 (2003): 5807–12.

Mohyeddin-Bonab, Mandana, Mohamad-Reza Mohamad-Hassani, Kamran Alimoghaddam, et al. "Autologous *In Vitro* Expanded Mesenchymal Stem Cell Therapy for Human Old Myocardial Infarction." *Archives of Iranian Medicine* 10, no. 4 (2007): 467–73.

Moise Jr., Kenneth J. "What to Tell Patients About Banking Cord Blood Stem Cells." *Contemporary OB/GYN* 51 (2006): 42–52.

Mombaerts, Peter. "Therapeutic Cloning in the Mouse." *Proceedings of the National Academy of Sciences USA* 100 (2003): 11924–5.

Moore, Keith L., and T. Vid N. Persuad. *The Developing Human: Clinically Oriented Embryology*. 8th ed. Philadelphia: W.B. Saunders Publishing Co., 2008.

Mothe, Andrea J., and Charles H. Tator. "Transplanted Neural Stem/Progenitor Cells Generate Myelinating Oligodendrocytes and Schwann Cells in Spinal Cord Demyelination and Dysmyelination." *Experimental Neurology* 213 (2008): 176–90.

Munné, Santiago. "Chromosome Abnormalities and their Relationship to Morphology and Development of Human Embryos." *Reproductive BioMedicine Online* 12, no. 2 (2005): 234–53.

Munsie, Megan J., Anna E. Michalska, Carmel M. O'Brien, Alan O. Trounson, Martin F. Pera, and Peter S. Mountford. "Isolation of Pluripotent Embryonic Stem Cells from Reprogrammed Adult Mouse Somatic Cell Nuclei." *Current Biology* 10, no. 6 (2000): 989–92.

Murphy, Jaime, Ross Summer, and Alan Fine. "Stem Cells in Airway Smooth Muscle: State of the Art." *Proceedings of the American Thoracic Society* 5, no. 11 (2008): 11–14.

Murphy, Mark J., Anne Wilson, and Andreas Trumpp. "More Than Just Proliferation: Myc Function in Stem Cells." *Trends in Cell Biology* 15 no. 3 (2005): 128–37.

Murphy, Sean V., and Anthony Atala. "Amniotic Fluid and Placental Membranes: Unexpected Sources of Highly Multipotent Cells." *Seminars in Reproductive Medicine* 31 (2013): 62–8.

Murphy, Sean, Rebecca Lim, Hayley Dickinson, Rutu Acharya, Sharina Rosli, Graham Jenkin, and Euan Wallace. "Human Amnion Epithelial Cells Prevent Bleomycin-Induced Lung Injury and Preserve Lung Function." *Cell Transplantation* 20 (2011): 909–23.

Murry, Charles E., Mark H. Soonpaa, Hans Reinecke, et al. "Haematopoietic Stem Cells Do Not Transdifferentiate into Cardiac Myocytes in Myocardial Infarcts." *Nature* 428 (2004): 664–8.

Nachtigall, Robert D., Gay Becker, Carrie Friese, Anneliese Butler, and Kirstin MacDougall. "Parent's Conceptualization of Their Frozen Embryos Complicates the Disposition Decision." *Fertility and Sterility* 84, no. 2 (2005): 431–4.

Naik, Gautam. "Scientists Create Stem-Cell Line." *Wall Street Journal. Jan.* 11, 2008.

Nakagawa, Masato, Michiyo Koyanagi, Koji Tanabe, et al. "Generation of Induced Pluripotent Stem Cells without Myc from Mouse and Human Fibroblasts." *Nature Biotechnology* 26 (2008): 101–6.

Nakajima, Hideaki, Kenzo Uchida, Alexander Rodriguez Guerrero, et al. "Transplantation of Mesenchymal Stem Cells Promotes the Alternative Pathway of Macrophage Activation and Functional Recovery after Spinal Cord Injury." *Journal of Neurotrauma* (2012). doi:10.1089/neu.2011.2109.

Nagy, A., A. Paldi, L. Dezso, L. Vargas, and A. Magyar. "Prenatal Fate of Parthenogenetic Cells in Mouse Aggregation Chimeras." *Development* 101 (1987): 67–71.

National Bioethics Advisory Commission. *Ethical Issues in Human Stem Cell Research*. Rockville, MD: National Bioethics Advisory Commission, 2000.

Neiva, Kathleen, Yan-Xi Sun and Russell S. Taichman. "The Role of Osteoblasts in Regulating Hematopoietic Stem Cell Activity and Tumor Metastasis." *Brazilian Journal of Medical and Biological Research* 38 (2005): 1449–54.

Nelson, Timothy J., Almudena Martinez-Fernandez, Satsuki Yamada, Carmen Perez-Terzic, Yasuhiro Ikeda, and Andre Terzic. "Repair of Acute Myocardial Infarction with iPS Induced by Human Stemness Factors." *Circulation* 120 (2009): 408–16.

Nelson, Timothy J., Zhi-Dong Ge, Jordan Van Orman, et al. "Improved Cardiac Function in Infarcted Mice after Treatment with Pluripotent Embryonic Stem Cells." *The Anatomical Record Part A: Discoveries in Molecular, Cellular, and Evolutionary Biology* 288A, no. 11 (2006): 1216–24.

Newton, C. R., J. Fisher, V. Feyles, F. Tekpetey, L. Hughes and D. Isacsson. "Changes in Patient Preferences in the Disposal of Cryopreserved Embryos." *Human Reproduction* 22, no. 12 (2007): 3124–8.

Nistor, Gabriel I., Minodora O. Totoiu, Nadia Haque, Melissa K. Carpenter, and Hans S. Keirstead. "Human Embryonic Stem Cells Differentiate into Oligodendrocytes in High Purity and Myelinate after Spinal Cord Transplantation." *Glia* 49 (2005): 385–96.

Niwa, Hitoshi. "How Is Pluripotency Determined and Maintained?" *Development* 134 (2007): 635–46.

Niwa, Hitoshi, Yayoi Toyooka, Daisuke Shimosato, et al. "Interaction between Oct3/4 and Cdx2 Determines Trophectoderm Differentiation." Cell 123, no. 5 (2005): 917–29.

Nomura, Tetsuya, Eishi Ashihara, Kento Tateishi, et al. "Therapeutic Potential of Stem/Progenitor Cells in Human Skeletal Muscle for Cardiovascular Regeneration." *Current Stem Cell Research & Therapy* 2, no. 4 (2008): 293–300.

Nori, Satoshi, Yohei Okada, Akimasa Yasuda, et al. "Grafted Human-Induced Pluripotent Stem-Cell–Derived Neurospheres Promote Motor Functional Recovery after Spinal Cord Injury in Mice." *Proceedings of the National Academy of Sciences USA* 108, no. 40 (2011): 16825–30.

Noyes, Nicole, Jeffrey Boldt, and Zsolt Peter Nagy. "Oocyte Cryopreservation: Is It Time to Remove Its Experimental Label?" *Journal of Assisted Reproduction and Genetics* 27 (2010): 69–74.

Nussbaum, Jeannette, Elina Minami, Michael A. Laflamme, et al. "Transplantation of Undifferentiated Murine Embryonic Stem Cells in the Heart: Teratoma Formation and Immune Response." *FASEB Journal* 21 (2007): 1345–57.

Nygren, Jens M., Stefan Jovinge, Martin Breitbach, et al. "Bone Marrow–Derived Hematopoietic Cells Generate Cardiomyocytes at a Low Frequency through Cell Fusion, but not Transdifferentiation." *Nature Medicine* 10 (2004): 494–501.

O'Donoghue, Keelin, and Jerry Chan. "Human Fetal Mesenchymal Stem Cells." *Current Stem Cell Research & Therapy* 1, no. 3 (2006): 371–86.

O'Rahilly, Ronan, and Fabiola Muller. *Human Embryology and Teratology*. 2nd ed. New York: Wiley-Liss, 1996.

Ogonuki, Narumi, Kimiko Inoue, Yoshie Yamamoto, et al. "Early Death of Mice Cloned from Somatic Cells." *Nature Genetics* 30 (2002): 253–4.

Oh, Hidemasa, Steven B. Bradfute, Teresa D. Gallardo, et al. "Cardiac Progenitor Cells from Adult Myocardium: Homing, Differentiation, and Fusion after Infarction." *Proceedings of the National Academy of Sciences USA* 100, no. 21 (2003): 12313–8.

Okada, Masaho, Thomas R. Payne, Bo Zheng, et al. "Myogenic Endothelial Cells Purified from Human Skeletal Muscle Improve Cardiac Function after Transplantation into Infarcted Myocardium." *Journal of the American College of Cardiology* 52, no. 23 (2008): 186980.

Okada, Seiji, Masaya Nakamura, Hiroyuki Katoh, et al. "Conditional Ablation of *Stat*3 or *Socs*3 Discloses a Dual Role for Reactive Astrocytes after Spinal Cord Injury." *Nature Medicine* 12 (2006): 829–34.

Okita, Keisuke, Tomoko Ichisaka, and Shinya Yamanaka. "Generation of Germline-Competent Induced Pluripotent Stem Cells." *Nature* 448 (2007): 313–7.

Okumura, Tomoyuki, Sophie S. W. Wang, Shigeo Takaishi, et al. "Identification of a Bone Marrow-Derived Mesenchymal Progenitor Cell Subset that Can Contribute to the Gastric Epithelium." *Laboratory Investigation* 89 (2009): 1410–22.

Okura, Hanayuki, Akifumi Matsuyama, Chun-Man Lee, et al. "Cardiomyoblast-like Cells Differentiated from Human Adipose Tissue-Derived Mesenchymal Stem Cells Improve Left Ventricular Dysfunction and Survival in a Rat Myocardial Infarction Model." *Tissue Engineering Part C: Methods* 16, no. 3 (2010): 417–25.

Ombelet, Willem, Petra De Sutter, Josiane Van der Elst, and Guy Martens. "Multiple Gestation and Infertility Treatment: Registration, Reflection and Reaction—The Belgian Project." *Human Reproduction Update* 11, no. 1 (2005): 3–14.

Opie, Lionel H., Patrick J, Commerford, Bernard J. Gersh, and Marc A. Pfeffer. "Controversies in Ventricular Remodeling." *Lancet* 367, no. 9507 (2006): 356–67.

Orlic, Donald, and Richard O. Cannon III. "Hematopoietic Stem Cells for Myocardial Regeneration" In *Stem Cells and Myocardial Regeneration*, edited by Marc S. Penn, 9–27. Totowa, NJ: Humana Press, 2007.

Orlic, Donald, Jan Kajstura, Stefano Chimenti, et al. "Bone Marrow Cells Regenerate Infarcted Myocardium." *Nature* 410 (2001): 701–5.

Orlic, Donald, Jan Kajstura, Stefano Chimenti, et al. "Mobilized Bone Marrow Cells Repair the Infarcted Heart, Improving Function and Survival." *Proceedings of the National Academy of Sciences USA* 98, no. 18 (2001): 10344–9.

Orr, Robert D., and Christopher C. Hook. "Stem Cell Research: Magical Promise v. Moral Peril." *Yale Journal of Health Policy, Law and Ethics* 2, no. 1 (2001): 189–99.

Oshima, Hideki, Thomas R. Payne, Kenneth L. Urish, et al. "Differential Myocardial Infarct Repair with Muscle Stem Cells Compared to Myoblasts." *Molecular Therapy* 12, no. 6 (2005): 113041.

Osiris Therapeutics, Inc. http://www.osiristx.com.

Paldi, A., A. Nagy, M. Markkula, I. Barna, and L. Dezso. "Postnatal Development of Parthenogenetic in Equilibrium with Fertilized Mouse Aggregation Chimeras." *Development* 105 (1989): 115–8.

Palmieri, C., P. Loi, G. Ptak., and L. Della Salda. "A Review of the Pathology of Abnormal Placentae of Somatic Cell Nuclear Transfer Clone Pregnancies in Cattle, Sheep, and Mice." *Veterinary Pathology* 45 (2008): 865–80.

Pan, Hung-Chuan, Fu-Chou Cheng, Shu-Zhen Lai, Dar-Yu Yang, Yeou-Chih Wang, and Maw-Sheng Lee. "Enhanced Regeneration in Spinal Cord Injury by Concomitant Treatment with Granulocyte Colony-Stimulating Factor and Neuronal Stem Cells." *Journal of Clinical Neuroscience* 15 (2008): 656–64.

Papapetrou, Eirini P., and Michel Sadelain. "Generation of Transgene-Free Human Induced Pluripotent Stem Cells With an Excisable Single Polycistronic Vector." *Nature Protocols* 6 (2011): 1251–73.

Park, Jin Hoon, Dae Yul Kim, Inn Young Sung, et al. "Long-Term Results of Spinal Cord Injury Therapy using Mesenchymal Stem Cells Derived from Bone Marrow in Humans." *Neurosurgery* (2012). doi: 10.1227/NEU.0b013e31824387f.

Park, In-Hyun, Natasha Arora, Hongguang Huo, et al. "Disease-Specific Induced Pluripotent Stem (iPS) Cells." *Cell* 134, no. 5 (2008): 877–86.

Parolini, Ornella, Francesco Alviano, Gian Paolo Bagnara, et al. "Concise Review: Isolation and Characterization of Cells from Human Term Placenta: Outcome of the First International Workshop on Placenta Derived Stem Cells." *Stem Cells* 26, no. 2 (2008): 300–11.

Parr, Ann M., Charles H. Tator, and Armand Keating. "Bone Marrow-Derived Mesenchymal Stromal Cells for the Repair of Central Nervous System Injury." *Bone Marrow Transplantation* 40 (2007): 609–19.

Passier, Robert, Linda W. van Laake, and Christine L. Mummery. "Stem-Cell-Based Therapy and Lessons from the Heart." *Nature* 453 (2008): 322–9.

Payne, John F., Douglas J. Raburn, Grace M. Couchman, Thomas M. Price, Margaret G. Jamison, and David K. Walmer. "Relationship Between Pre-Embryo Pronuclear Morphology (Zygote Score) and Standard Day 2 or 3 Embryo Morphology with Regard to Assisted Reproductive Technique Outcomes" *Fertility and Sterility* 84, no. 4 (2005): 900–9.

Pearse, Damien D., Andre R. Sanchez, Francisco C. Pereira, et al. "Transplantation of Schwann Cells and/or Olfactory Ensheathing Glia into the Contused Spinal Cord: Survival, Migration, Axon Association, and Functional Recovery." *Glia* 55, no. 9 (2007): 976–1000.

Pearson, Helen. "Health Effects of Egg Donation May Take Decades to Emerge." *Nature* 442 (2006): 607–8.

Pera, Mark F. "Stem Cells: The Dark Side of Induced Pluripotency." *Nature* 471 (2011): 46–7.

Percer, Beth. "Umbilical Cord Blood Banking: Helping Parents Make Informed Choices." *Nursing for Women's Health* 13, no. 3 (2009): 217–23.

Pergament E., M. Fiddler, N. Cho, D. Johnson, and W.J. Holmgren. "Sexual Differentiation and Preimplantation Cell Growth." *Human Reproduction* 91 (1994): 1730–2.

Perin, L., S. Giuliani, D. Jin, et al. "Renal Differentiation of Amniotic Fluid Stem Cells." *Cell Proliferation* 40, no. 6 (2007): 936–48.

Perin, Laura, Sargis Sedrakyan, Stefano Giuliani, et al. "Protective Effect of Human Amniotic Fluid Stem Cells in an Immunodeficient Mouse Model of Acute Tubular Necrosis." *PLoS One* 5, no. 2 (2010): e9357. doi:10.1371/journal.pone.0009357.

Perlman, David. "Embryonic Research Brings Great Hopes: Cell Studies May Yield Variety of Cures within Next 10 Years." *San Francisco Chronicle*, November 6, 1998.

Peters, Ted. *The Stem Cell Debate*. Minneapolis: Fortress Press, 2007.

Peterson, James C. "The Ethics of the ANT Proposal to Obtain Embryo-Type Stem Cells." *Perspectives on Science and the Christian Faith* 58, no. 4 (2006): 294–302.

Philo of Alexandria. *The Works of Philo*. Translated by C. D. Yonge. Peabody, MA: Hendrickson Publishers, 1993.

Phinney, Donald G., and Darwin J. Prockop. "Concise Review: Mesenchymal Stem/Multipotent Stromal Cells: The State of Transdifferentiation and Modes of Tissue Repair—Current Views." *Stem Cells* 25 (2007): 2896–902.

Pietersma, Albert, and Benjamin G. Wright, eds. *A New English Translation of the Septuagint*. New York: Oxford University Press, 2007. Kindle edition.

Pittenger, Mark F., Alastair M. Mackay, Stephen C. Beck, et al. "Multilineage Potential of Adult Human Mesenchymal Stem Cells." *Science* 284 (1999): 143–7.

Plemel, Jason R., Andrew Chojnacki, Joseph S. Sparling, et al. "Platelet-Derived Growth Factor-Responsive Neural Precursors Give Rise to Myelinating Oligodendrocytes after Transplantation into the Spinal Cords of Contused Rats and Dysmyelinated Mice." *Glia* 59, no. 12 (2011): 1891–1910.

Polejaeva, Irina A., Shu-Hung Chen, Todd D. Vaught, et al. "Cloned Pigs Produced by Nuclear Transfer from Adult Somatic Cells." *Nature* 407 (2000): 86–90.

Poon, Ellen, Chi-wing Kong, and Ronald A. Li. "Human Pluripotent Stem Cell-Based Approaches for Myocardial Repair: From the Electrophysiological Perspective." *Molecular Pharmaceutics* 8 (2011): 1495–1504.

Practice Committee of the American Society for Reproductive Medicine. "Ovarian Hyperstimulation Syndrome." *Fertility and Sterility* 80 (2003): 1309–14.

Prentice, David A. "Current Science of Regenerative Medicine with Stem Cells." *Journal of Investigative Medicine* 54, no. 1 (2006): 33–7.

President's Council on Bioethics. *Reproduction and Responsibility: Regulating new Biotechnologies*. Washington D.C.; March 2004. http://bioethics.georgetown.edu/pcbe/reports/reproductionandresponsibility/chapter2.html.

President's Council on Bioethics. *Stem Cell Research: Recent Scientific and Clinical Development*. Session 3. July 24, 2003. http://bioethics.georgetown.edu/pcbe/transcripts/july03/session3.html

Prigozhina, Tatyana B., Sofia Khitrin, Gregory Elkin, Osnat Eizik, Shoshana Morecki, and Shimon Slavin. "Mesenchymal Stromal Cells Lose Their Immunosuppressive Potential after Allotransplantation." *Experimental Hematology* 36, no. 10 (2008): 1370–6.

Prockop, Darwin J. "Marrow Stromal Cells as Stem Cells for Nonhematopoietic Tissues." *Science* 276 (1997): 71–4.

Qiao, Hui, Hualei Zhang, Satoshi Yamanaka, et al. "Long-Term Improvement in Postinfarct Left Ventricular Global and Regional Contractile Function is Mediated by Embryonic Stem Cell-Derived Cardiomyocytes." *Circulation* 4 (2011): 33–41.

Quevedo, Henry C., Konstantinos E. Hatzistergos, Behzad N. Oskouei, et al. "Allogeneic Mesenchymal Stem Cells Restore Cardiac Function in Chronic Ischemic Cardiomyopathy Via Trilineage Differentiating Capacity." *Proceedings of the National Academy of Sciences, USA* 106, no. 33 (2009): 14022–7.

Ra, Jeong Chan, Sung Keun Kang, Il Seob Shin, et al. "Stem Cell Treatment for Patients with Autoimmune Disease by Systemic Infusion of Culture-Expanded Autologous Adipose Tissue Derived Mesenchymal Stem Cells." *Journal of Translational Medicine* 9 (2011): 181.

Rabusin, M., M. Andolina, N. Maximova, et al. "Immunoablation Followed by Autologous Hematopoietic Stem Cell Infusion for the Treatment of Severe Autoimmune Disease." supplement, *Haematologica* 85, no. S11 (2000): 81–5.

Rachakatla, Raja Shekar, F. Marini, M. L. Weiss, M. Tamura, and D. Troyer. "Development of Human Umbilical Cord Matrix Stem Cell-Based Gene Therapy for Experimental Lung Tumors." *Cancer Gene Therapy* 14 (2007): 828–35.

Rachakatla, Raja Shekar, Marla M. Pyle, Rie Ayuzawa, et al. "Combination Treatment of Human Umbilical Cord Matrix Stem Cell-Based Interferon-Beta Gene Therapy and 5-Fluorouracil Significantly Reduces Growth of Metastatic Human Breast Cancer in SCID Mouse Lungs." *Cancer Investigation*, 26, no. 7 (2008): 662–70.

Rajala, Kristiina, Bettina Lindroos, Samer M. Hussein, et al. "A Defined and Xeno-Free Culture Method Enabling the Establishment of Clinical-Grade Human Embryonic, Induced Pluripotent and Adipose Stem Cells." *PLoS One* 5, no. 4 (2010): e10246. doi:10.1371/journal.pone.0010246.

Rajala, Kristiina, Mari Pekkanen-Mattila, and Katriina Aalto-Setälä. "Cardiac Differentiation of Pluripotent Stem Cells." *Stem Cells International* 2011 (2011): 1–12. doi:10.4061/2011/383709.

Ralston, Amy and Janet Rossant. "Genetic Regulation of Stem Cell Origins in the Mouse Embryo." *Clinical Genetics* 68, no. 2 (2005): 106–12.

Ramón-Cueto, Almudena, M. Isabel Cordero, Fernando F. Santos-Benito, and Jesús Avila. "Functional Recovery of Paraplegic Rats and Motor Axon Regeneration in Their Spinal Cords by Olfactory Ensheathing Glia." *Neuron* 25 (2000): 425–35.

Ratajczak, J., E. Zuba-Surma, E. Paczkowska, M. Kucia, P. Nowacki, and M. Z. Rataczak. "Stem Cells for Neural Regeneration—A Potential Application of Very Small Embryonic-Like Stem Cells." Journal of Physiological Pharmacology 61, no. 1 (2011): 3–12.

Ratajczak, Mariusz Z., Ewa K. Zuba-Surma, Boguslaw Machalinski, and Magdalena Kucia. "Bone-Marrow-Derived Stem Cells—Our Key to Longevity?" *Journal of Applied Genetics* 48, no. 4 (2007): 307–19.

Ratajczak, Mariusz Z., Rui Liu, Janina Ratajczak, Magda Kucia, and Dong-Myung Shin. "The Role of Pluripotent Embryonic-Like Stem Cells Residing in Adult Tissues in Regeneration and Longevity." *Differentiation* 81, no. 3 (2011): 153–61.

Rehni, Ashish K., Nirmal Singh, Amteshwar S. Jaggi, and Manjeet Singh. "Amniotic Fluid Derived Stem Cells Ameliorate Focal Cerebral Ischaemia-Reperfusion Injury Induced Behavioural Deficits in Mice." *Behavioural Brain Research* 183, no. 1 (2007): 95–100.

Renard, Jean-Paul, Sylvie Chastant, Patrick Chesné et al. "Lymphoid Hypoplasia and Somatic Cloning." *Lancet* 353 (1999): 1489–91.

Repas-Humpe, L. M., A. Humpe, R. Lynen, et al. "A Dispermic Chimerism in a 2-Year-Old Caucasian Boy." *Annuals of Hematology* 78 (1999): 431–4.

Revazova, E. S., N. A. Turovets, O. D. Kochetkova, et al. "Patient-Specific Stem Cell Lines Derived from Human Parthenogenetic Blastocysts." *Cloning and Stem Cells* 9, no. 3 (2007): 432–49.

Revazova, E. S., N. A. Turovets, O. D. Kochetkova, et al. "HLA Homologous Stem Cell Lines Derived from Human Parthenogenetic Blastocysts." *Cloning and Stem Cells* 10, no. 1 (2008): 11–24.

Revelli, Alberto, Emanuela Molinari, Francesca Salvagno, Luisa Delle Piane, Elisabetta Dolfin, and Simona Ochetti. "Oocyte Cryostorage to Preserve Fertility in Oncological Patients." *Obstetrics and Gynecology International* 2012 (2012): 525896. http://www.ncbi.nlm.nih.gov/pmc/articles/PMC3265124/pdf/OGI2012-525896.pdf.

Rhind, Susan M., Jane E. Taylor, Paul A. De Sousa, Tim J. King, Michelle McGarry, and Ian Wilmut. "Human Cloning: Can It Be Made Safe?" *Nature Reviews Genetics* 4 (2003): 855–64.

Rhind, Susan M., Tim J. King, Linda M. Harkness, et al. "Cloned Lambs—Lessons from Pathology." *Nature Biotechnology* 21 (2003): 744–5.

Richter, Miranda W., and A. Jane Roskams. "Olfactory Ensheathing Cell Transplantation Following Spinal Cord Injury: Hype or Hope?" *Experimental Neurology* 209, no. 2 (2008): 353–67.

Rideout, William M., Konrad Hochedlinger, Michael Kyba, George Q. Daley, and Rudolf Jaenisch. "Correction of a Genetic Defect by Nuclear Transplantation and Combined Cell and Gene Therapy." *Cell* 109, no. 1 (2002): 17–27.

Roberts, Alexander, James Donaldson, and A. Cleveland Coxe, eds. *Ante-Nicene Fathers*. 10 vols. Grand Rapids, MI: Wm. B. Eerdmans Publishing Co., 1950–51. http://www.ccel.org/fathers.html.

Robinson, Richard. "Another Set Back for Fetal Transplants for Parkinson's Disease." *Lancet Neurology* 2, no. 2 (2003): 69.

Robson, Sam, Stella Pelengaris, and Michael Khan. "c-Myc and Downstream Targets in the Pathogenesis and Treatment of Cancer." *Recent Patents on Anticancer Drug Discovery* 1, no. 3 (2006): 305–26.

Rocha, Vanderson, and Franco Locatelli. "Searching for Alternative Hematopoietic Stem Cell Donors for Pediatric Patients." *Bone Marrow Transplantion* 41 (2008): 83–93.

Rocha, Vanderson, Jacqueline Cornish, Eric L. Sievers, et al. "Comparison of Outcomes of Unrelated Bone Marrow and Umbilical Cord Blood Transplants in Children with Acute Leukemia." *Blood* 97, no. 10 (2001): 2962–71.

Rocha, Vanderson, Myriam Labopin, Guillermo Sanz, et al. "Transplants of Umbilical-Cord Blood or Bone Marrow from Unrelated Donors in Adults with Acute Leukemia." *New England Journal of Medicine* 351 (2004): 2276–85.

Roell, Wilhelm, Thorsten Lewalter, Philipp Sasse, et al. "Engraftment of Connexin 43-Expressing Cells Prevents Post-Infarct Arrhythmia." *Nature* 450 (2007): 819–24.

Rota, Marcello, Jan Kajstura, Toru Hosoda, et al. "Bone Marrow Cells Adopt the Cardiomyogenic Fate *In Vivo*." *Proceedings of the National Academy of Sciences USA* 104, no. 45 (2007): 17783–8.

Roufosse, Candice A., N. C. Direkze, W. R. Otto, and N. A. Wright. "Circulating Mesenchymal Stem Cells." *International Journal of Biochemistry and Cell Biology* 36, no. 4 (2004):585–97.

Rowland Hogue, Carol J. "Successful Assisted Reproduction Technology: The Beauty of One." *Obstetrics and Gynecology* 100, no. 5 (2002): 1017–9.

Sakai, Kiyoshi, Akihito Yamamoto, Kohki Matsubara, et al. "Human Dental Pulp-Derived Stem Cells Promote Locomotor Recovery after Complete Transection of the Rat Spinal Cord by Multiple Neuro-Regenerative Mechanisms." *Journal of Clinical Investigation* 122, no. 1 (2012): 80–90.

Sakuragawa, Norio, Kenichi Kakinuma, Aiko Kikuchi, et al. "Human Amnion Mesenchyme Cells Express Phenotypes of Neuroglial Progenitor Cells." *Journal of Neuroscience Research* 78, no. 2 (2004): 208–14.

Sandel, Michael. "Embryo Ethics—The Moral Logic of Stem-Cell Research" *New England Journal of Medicine* 351 (2004): 207–9.

Sanmano, Borisut, Masayuki Mizoguchi, Yasushi Suga, Shigaku Ikeda, and Hideoki Ogawa. "Engraftment of Umbilical Cord Epithelial Cells in Athymic Mice: In an Attempt to Improve Reconstructed Skin Equivalents Used as Epithelial Composite." *Journal of Dermatological Science* 37 (2005): 29–39.

Sarugaser, Rahul, David Lickorish, Dolores Baksh, M. Morris Hosseini, and John E. Davies. "Human Umbilical Cord Perivascular (HUCPV) Cells: A Source of Mesenchymal Progenitors." *Stem Cells* 23, no. 2 (2005): 220–9.

Savulescu, Julian. "Should We Clone Human Beings? Cloning as a Source of Tissue for Transplantation." *Journal of Medical Ethics* 25 (1999): 87–95.

Schaff, Philip, ed. *Nicene and Post-Nicene Fathers: First Series*. 14 vols. Grand Rapids, MI: Wm. B. Eerdmans Publishing Co., 1956. http://www.ccel.org/fathers.html.

———, ed. *Nicene and Post-Nicene Fathers: Second Series*. 14 vols. Grand Rapids, MI: Wm. B. Eerdmans Publishing Co., 1952–57. http://www.ccel.org/fathers.html.

Schira, Jessica, Marcia Gasis, Veronica Estrada, et al. "Significant Clinical, Neuropathological and Behavioural Recovery from Acute Spinal Cord Trauma by Transplantation of a Well-Defined Somatic Stem Cell from Human Umbilical Cord Blood." *Brain* 135, no. 2 (2012): 431–46.

Schmidt, Alvin J. *How Christianity Changed the World*. Grand Rapids, MI: Zondervan, 2004.

Schmidt, Dörthe, Josef Achermann, Bernhard Odermatt, et al. "Prenatally Fabricated Autologous Human Living Heart Valves Based on Amniotic Fluid Derived Progenitor Cells as Single Cell Source." *Circulation* 116 (2007): I-64–70.

Schoemans, H., K. Theunissen, J. Maertens, M. Boogaerts, C. Verfaillie, and J. Wagner. "Adult Umbilical Cord Blood Transplantation: A Comprehensive Review." *Bone Marrow Transplantation* 38, no. 2 (2006): 83–93.

Schoenwolf, Gary C., Steven B. Bleyl, Philip R. Brauer, and Philippa H. Francis-West. *Larsen's Human Embryology*, 4th ed. Philadelphia: Churchill Livingstone Elsevier, 2009.

Schwartz, Steven D., Jean-Pierre Hubschman, Gad Heilwell, et al. "Embryonic Stem Cell Trials for Macular Degeneration: A Preliminary Report." *Lancet* 379 (2012): 713–20.

Sell, Stewart. "Stem Cells: What are They? Where do They Come From? Why are They Here? When do They go Wrong? Where Are They Going?" In *Stem Cells Handbook*, edited by Stewart Sell, 1–18. Totowa, NJ: Humana Press, 2003.

Sha, Ky. "A Mechanistic View of Genomic Imprinting." *Annual Review of Genomics and Human Genetics* 9 (2008): 197–216.

Shake, Jay G., Peter J. Gruber, William A. Baumgartner, et al. "Mesenchymal Stem Cell Implantation in a Swine Myocardial Infarct Model: Engraftment and Functional Effects." *Annals of Thoracic Surgery* 73 (2002): 1919–26.

Shannon, Thomas, and Allan Wolter. "Reflections on the Moral Status of the Pre-Embryo." *Theological Studies* 51 (1990): 603–26.

Sharp, Jason, and Hans S. Keirstead. "Therapeutic Applications of Oligodendrocyte Precursors Derived from Human Embryonic Stem Cells." *Current Opinion in Biotechnology* 18 (2007): 434–40.

Shen, Jie-fei, Atsunori Sugawara, Joe Yamashita, Hideo Ogura, and Soh Sato. "Dedifferentiated Fat Cells: An Alternative Source of Adult Multipotent Cells from the Adipose Tissues." *International Journal of Oral Science* 3 (2011): 117–124.

Sheng, Zhiyong, Xiaobing Fu, Sa Cai, et al. "Regeneration of Functional Sweat Gland-like Structures by Transplanted Differentiated Bone Marrow Mesenchymal Stem Cells." *Wound Repair and Regeneration* 17 (2009): 427–35.

Sherman, Warren. "Cell Therapy in the Cath Lab for Heart Failure: A Look at MyoCell Therapy and the SEISMIC Trial." *Cath Lab Digest* 16, no. 5 (2008): 1–11.

Shi, M., Z.-W. Liu and F.-S. Wang. "Immunomodulatory Propertiess and Therapeutic Application of Mesenchymal Stem Cells." *Clinical and Experimental Immunology* 164 (2011): 1–8.

Shiga, Kazuho, K. Hirose, Y. Sasae, and T. Nagai. "Production of Calves by Transfer of Nuclei from Cultured Somatic Cell Obtained from Japanese Black Bulls." *Theriogenology* 52 (1999): 527–35.

Shin, Taeyoung, Duane Kraemer, Jane Pryor, et al. "A Cat Cloned by Nuclear Transplantation." *Nature* 415 (2002): 859.

Shi-Xia, Xu, Tang Xian-Hua, and Tang Xiang-Feng. "Unrelated Umbilical Cord Blood Transplantation and Unrelated Bone Marrow Transplantation in Children with Hematological Disease: A Meta-Analysis." *Pediatric Transplantation* 13, no. 3 (2009): 278–84.

Shizuru, Judith A., Robert S. Negrin, and Irving L. Weissman. "Hematopoietic Stem and Progenitor Cells: Clinical and Preclinical Regeneration of the Hematolymphoid System." *Annual Review of Medicine* 56 (2005): 509–38.

Sideri, Anastasia, Nikolaos Neokleous, Philippe Brunet De La Grange, et al. "An Overview of the Progress on Double Umbilical Cord Blood Transplantation." *Haematologica* 96, no. 8 (2011): 1213–20.

Sidhu, Kuldip S. "New Approaches for the Generation of Induced Pluripotent Stem Cells." *Expert Opinion on Biological Therapy* 11, no. 5 (2011): 567–79.

Sills, Scott E., Michael J. Tucker, and Gianpiero D. Palermo. "Assisted Reproductive Technologies and Monozygous Twins: Implications for Future Study and Clinical Practice." *Twin Research* 4 (2000): 217–23.

Silver, Lee. *Remaking Eden: Cloning and Beyond in a Brave New World.* New York: Avon Books, 1997.

Siminiak, Tomasz, Dorota Fiszer, Olga Jerzykowska, et al. "Percutaneous Trans-coronary-Venous Transplantation of Autologous Skeletal Myoblasts in the Treatment of Post-infarction Myocardial Contractility Impairment: The POZNAN Trial." *European Heart Journal* 26, no. 12 (2005): 1188–95.

Siminiak, Tomasz, Ryszard Kalawski, Dorota Fiszer, et al. "Autologous Skeletal Myoblast Transplantation for the Treatment of Postinfarction Myocardial Injury: Phase I Clinical Study with 12 Months of Follow-Up." *American Heart Journal* 148, no. 3 (2004): 531–7.

Simpson, Joe Leigh. "Blastomeres and Stem Cells." *Nature* 444 (2006): 432–5.

Sinclair, Kevin D., T. G. McEvoy, E. K. Maxfield, et al. "Aberrant Fetal Growth and Development After In Vitro Culture of Sheep Zygotes." *Journal of Reproductive Fertility* 116, no. 1 (1999): 177–86.

Singla, Dinender K., Xilin Long, Carley Glass, Reetu D. Singla, and Binbin Yan. "Induced Pluripotent Stem (iPS) Cells Repair and Regenerate Infarcted Myocardium." *Molecular Pharmaceutics* 8, no. 5 (2011): 1573–81.

Slim, Rima and Amira Mehio. "The Genetics of Hydatiform Moles: New Lights on an Ancient Disease." *Clinical Genetics* 71 (2005): 25–34.

Smith, Barry, and Berit Brogaard. "Sixteen Days." *The Journal of Medicine and Philosophy* 28 (2003): 45–78.

Smith, Rachel Ruckdeschel, Lucio Barile, Hee Cheol Cho, et al. "Regenerative Potential of Cardiosphere-Derived Cells Expanded from Percutaneous Endomyocardial Biopsy Specimens." *Circulation* 115 (2007): 896–908.

Smith, Sadie L., Robin E. Everts, Li-Ying Sung, et al. "Gene Expression Profiling of Single Bovine Embryos Uncovers Significant Effects of In Vitro Maturation, Fertilization and Culture." *Molecular Reproduction and Development* 76, no. 1 (2008): 38–47.

Smith, Wesley J. *To The Source* (blog). http://www.tothesource.org/index.htm.

———. *Secondhand Smoke* (blog). http://www.nationalreview.com/human-exceptionalism.

Smits, Pieter C., Robert-Jan M. van Geuns, Don Poldermans, et al. "Catheter-Based Intramyocardial Injection of Autologous Skeletal Myoblasts as a Primary Treatment of Ischemic Heart Failure." *Journal of American College of Cardiology* 42, no. 12 (2003): 2063–9.

Soldner, Frank, Dirk Hockemeyer, Caroline Beard, et al. "Parkinson's Disease Patient-Derived Induced Pluripotent Stem Cells Free of Viral Reprogramming Factors." *Cell* 136, no. 5 (2009): 964–77.

Somers, Joanna, Craig Smith, Martyn Donnison, et al. "Gene Expression Profiling of Individual Bovine Nuclear Transfer Blastocysts." *Reproduction* 13, no. 1 (2006): 1073–84.

Someya, Yukio and Masao Koda. "Reduction of Cystic Cavity, Promotion of Axonal Regeneration and Sparing, and Functional Recovery with Transplanted Bone Marrow Stromal Cell–Derived Schwann Cells after Contusion Injury to the Adult Rat Spinal Cord." *Journal of Neurosurgery: Spine* 9, no. 6 (2008): 600–10.

Soncino Babylonian Talmud Niddah. Edited by Isidore Epstein. Translated by Israel W. Slotki. Teaneck, New Jersey: Talmudic Books, 2012. Kindle edition.

Song, Heesang, Byeong-Wook Song, Min-Ji Cha, In-Geol Choi, and Ki-Chul Hwang. "Modification of Mesenchymal Stem Cells for Cardiac Regeneration." *Expert Opinion on Biological Therapy* 10, no. 3 (2010): 309–19.

Soto-Gutierrez, Alejandro, Li Zhang, Chris Medberry, et al. "A Whole-Organ Regenerative Medicine Approach for Liver Replacement." *Tissue Engineering Part C: Methods* 17, no. 6 (2011): 677–86.

Sourvinos, George, Christos Tsatsanis, and Demetrios A. Spandidos. "Mechanisms of Retrovirus-Induced Oncogenesis." *Folia Biologia* 46, no. 6 (2000): 226–32.

Spar, Debora. "The Egg Trade – Making Sense of the Market for Human Oocytes." *New England Journal of Medicine* 365, no 13 (2007): 1289–91.

Srikanth, Garikipati V. N., Naresh K. Tripathy, and Soniya Nityanand. "Fetal Cardiac Mesenchymal Stem Cells Express Embryonal Markers and Exhibit Differentiation into Cells of All Three Germ Layers." *World Journal of Stem Cells* 5, no. 1 (2012): 26–33.

Stamm, Christof, Bernd Westphal, Hans-Dieter Kleine, et al. "Autologous Bone-Marrow Stem-Cell Transplantation for Myocardial Regeneration." *Lancet* 361, no. 9351 (2003): 45–6.

Stamm, Christof, Hans-Dieter Kleine, Yeong-Hoon Choi, et al. "Intramyocardial Delivery of CD133+ Bone Marrow Cells and Coronary Artery Bypass Grafting for Chronic Ischemic Heart Disease and Efficacy Studies." *Journal of Thoracic and Cardiovascular Surgery* 133, no. 3 (2007): 717–25.

Steigenga, Marc J., Frans M. Helmerhorst, Jurien De Koning, Ans M. I. Tijssen, Sebastiaan A. T. Ruinard, and Frietson Galis. "Evolutionary Conserved Structures as Indicators of Medical Risks: Increased Incidence of Cervical Ribs after Ovarian Hyperstimulation in Mice." *Animal Biology* 56, no. 1 (2006): 63–8.

Steinbrook, Robert. "Egg Donation and Human Embryonic Stem-Cell Research." *New England Journal of Medicine* 354, no. 4 (2006): 324–6.

Stephenson, Emma, Laureen Jacquet, Cristian Miere, et al. "Derivation and Propagation of Human Embryonic Stem Cell Lines from Frozen Embryos in an Animal Product–Free Environment." *Nature Protocols* 7 (2012): 1366–81.

Strain, Lisa, John C. S. Dean, Mark P. R. Hamilton, and David T. Bonthron. "A True Hermaphrodite Chimera Resulting from Embryo Amalgamation after in Vitro Fertilization." *New England Journal of Medicine* 338 (1998): 166–9.

Strelchenko, Nick, and Yury Verlinsky. "Embryonic Stem Cells from Morula." *Methods in Enzymology* 418 (2006):93–108.

Strumpf, Dan, Chai-An Mao, Yojiro Yamanaka, et al. "Cdx2 is Required for Correct Cell Fate Specification and Differentiation of Trophectoderm in the Mouse Blastocyst." *Development* 132 (2005): 2093–2102.

Sugawara, Atsushi, Brittany Sato, Elise Bal, Abby C. Collier, and Monika A. Ward. "Blastomere Removal from Cleavage-Stage Mouse Embryos Alters Steroid Metabolism During Pregnancy." *Biology of Reproduction* 87 (2012): 1–9.

Sugawara, Atsushi and Monika A. Ward. "Biopsy of Embryos Produced by In Vitro Fertilization Affects Development in C57BL/6 Mouse Strain." *Theriogenology* 79 (2013): 234–41.

Sun, Hongli, Kai Feng, Jiang Hu, Shay Soker, Anthony Atala, and Peter X. Ma. "Osteogenic Differentiation of Human Amniotic Fluid-Derived Stem Cells Induced by Bone Morphogenetic Protein-7 and Enhanced by Nanofibrous Scaffolds." *Biomaterials* 31, no. 6 (2010): 1133–9.

Swartz, Stephen D. *The Moral Question of Abortion*. Chicago: Loyola University Press, 1990.

Swijnenburg, Rutger-Jan, Masashi Tanaka, Hannes Vogel, et al. "Embryonic Stem Cell Immunogenicity Increases upon Differentiation after Transplantation into Ischemic Myocardium." supplement, *Circulation* 112, no. S1, (2005): I-166–72.

Sybert, Virginia P. "Hypomelanosis of Ito: A Description, not a Diagnosis." *Journal of Investigative Dermatology* 103 (1994): 141S–3S.

Syková, Eva, Pavla Jendelová, Lucia Urdzíková, Petr Lesný, and Aleš Hejčl. "Bone Marrow Stem Cells and Polymer Hydrogels – Two Strategies for Spinal Cord Injury Repair." *Cellular and Molecular Neurobiology* 26, no. 7–8 (2006): 1113–29.

Tabar, Viviane, Mark Tomishima, Georgia Panagiotakos, et al. "Therapeutic Cloning in Individual Parkinsonian Mice." *Nature Medicine* 14, no. 4 (2008): 379–81.

Takahashi, Kazutoshi, Koji Tanabe, Mari Ohnuki, et al. "Induction of Pluripotent Stem Cells from Adult Human Fibroblasts by Defined Factors." *Cell* 131 (2007): 861–72.

Takami, Toshihiro, Martin Oudega, Margaret L. Bates, Patrick M. Wood, Naomi Kleitman, and Mary Bartlett Bunge. "Schwann Cell But Not Olfactory Ensheathing Glia Transplants Improve Hindlimb Locomotor Performance in the Moderately Contused Adult Rat Thoracic Spinal Cord." *Journal of Neuroscience* 22, no. 15 (2002): 6670–81.

Tamashiro, Kellie L.K., Teruhiko Wakayama, Hidenori Akutsu, et al. "Cloned Mice Have an Obese Phenotype Not Transmitted to Their Offspring." *Nature Medicine* 8 (2002): 262–7.

Tanaka, Satoshi, Mayumi Oda, Yasushi Toyoshima, et al. "Placentomegaly in Cloned Mouse Concepti Caused by Expansion of the Spongiotrophoblast Layer." *Biology of Reproduction* 65 (2001): 1813–21.

Tang, Junming, Qiyang Xie, Guodong Pan, Jianing Wang, and Mingjiang Wang. "Mesenchymal Stem Cells Participate in Angiogenesis and Improve Heart Function in Rat Model of Myocardial Ischemia with Reperfusion." *European Journal of Cardio-Thoracic Surgery* 30, no. 2 (2006): 353–61.

Tannenbaum, Shelly E., Tikva T. Turetsky, Orna Singer, et al. "Derivation of Xeno-Free and GMP-Grade Human Embryonic Stem Cells – Platforms for Future Clinical Applications." *PLoS One* 7, no. 6 (2012): e35325.

Tarin, J. J., J. Conaghan, R. M. L. Winston, and A. H. Handyside. "Human Embryo Biopsy on the 2nd Day After Insemination for Preimplantation Diagnosis — Removal of a Quarter of Embryo Retards Cleavage." *Fertility and Sterility* 58 (1992): 970–6.

Taura, Daisuke, Michio Noguchi, Masakatsu Sone, et al. "Adipogenic Differentiation of Human Induced Pluripotent Stem Cells: Comparison with That of Human Embryonic Stem Cells." *FEBS Letters* 583, no. 6 (2009): 1029–33.

Telles, Paloma Dias, Maria Aparecida de Andrade Moreira Machado, Vivien Thiemy Sakai, and Jacques Eduardo Nör. "Pulp Tissue from Primary Teeth: New Source of Stem Cells." *Journal of Applied Oral Science* 19, no. 3 (2011): 189–94.

Theise, Neil D., Manjunath Nimmakayalu, Rebekah Gardner, et al. "Liver from Bone Marrow in Humans." *Hepatology* 32 (2000): 11–16.

Thiele, J., E. Varus, C. Wickenhauser, H. M. Kvasnicka, K. A. Metz, and D. W. Beelen. "Liver from Bone Marrow in Humans." *Hepatology* 32 (2000): 11–16.

Thompson., Charis. "Why We Should, In Fact, Pay for Egg Donation." *Regenerative Medicine* 2, no. 2 (2007): 203–9.

Thomson, James A., Jennifer Kalishman, Thaddeus G. Golos, et al. "Isolation of a Primate Embryonic Stem Cell Line." *Proceedings of the National Academy of Sciences USA* 92 (1995): 7844–8.

Thomson, James A., Jennifer Kalishman, Thaddeus G. Golos, Maureen Durning, Charles P. Harris, and John P. Hearn, "Pluripotent Cell Lines Derived from Common Marmoset *(Callithrix jacchus)* Blastocysts." *Biology of Reproduction* 55 (1996): 254–9.

Thomson, James A., Joseph Itskovitz-Eldor, Sander S. Shapiro, et al. "Embryonic Stem Cells Lines Derived from Human Blastocysts." *Science* 282 (1998): 1145–7.

Toma, Catalin, Mark F. Pittenger, Kevin S. Cahill, Barry J. Byrne, and Paul D. Kessler. "Human Mesenchymal Stem Cells Differentiate to a Cardiomyocyte Phenotype in the Adult Murine Heart." *Circulation* 105 (2002): 93–8.

Totoiu, Minodora O., and Hans S. Keirstead. "Spinal Cord Injury Is Accompanied by Chronic Progressive Demyelination." *Journal of Comparative Neurology* 486 (2005): 373–83.

Traber, Peter G., and Debra G. Stilberg. "Intestine-Specific Gene Transcription." *Annual Reviews of Physiology* 58 (1996): 275–97.

Trobridge, Grant D. "Genotoxicity of Retroviral Hematopoietic Stem Cell Gene Therapy." *Expert Opinion on Biological Therapy* 11, no. 5 (2011): 581–93.

Trounson, Alan. "Stem Cells, Plasticity and Cancer - Uncomfortable Bed Fellows." *Development* 131, no. 12 (2004): 2763–8.

Tsuji, Osahiko, Kyoko Miura, Yohei Okada, et al. "Therapeutic Potential of Appropriately Evaluated Safe-Induced Pluripotent Stem Cells for Spinal Cord Injury." *Proceedings of the National Academy of Sciences USA* 107, no. 28 (2010): 12704–9.

Tyler Moore, Mary. *Testimony on Behalf of the Juvenile Diabetes Foundation before the Senate Appropriations Subcommittee on Labor, Health and Human Services and Education*, September 14, 2000.

Tyndall, Alan, and Antonio Uccelli. "Multipotent Mesenchymal Stromal Cells for Autoimmune Diseases: Teaching New Dogs Old Tricks." *Bone Marrow Transplantation* 43 (2009): 821–8.

Uccelli, Antonio, Alice Laroni, and Mark S Freedman. "Mesenchymal Stem Cells for the Treatment of Multiple Sclerosis and Other Neurological Diseases." *Lancet Neurology* 10, no. 7 (2011): 649–56.

Ugajin, Tomohisa, Yukihiro Terada, Hisataka Hasegawa, Clarissa L. Velayo, Hiroshi Nabeshima, and Nobuo Yaegashi. "Aberrant Behavior of Mouse Embryo Development After Blastomere Biopsy as Observed Through Time-Lapse Cinematography." *Fertility and Sterility* 93 (2010): 2723–8.

Undale, Anita H., Jennifer J. Westendorf, Michael J. Yaszemski, and Sundeep Khosla. "Mesenchymal Stem Cells for Bone Repair and Metabolic Bone Diseases." *Mayo Clinic Proceedings* 84, no. 10 (2009): 9893–902.

Urbich, Carmen, and Stephanie Dimmeler. "Endothelial Progenitor Cells: Characterization and Role in Vascular Biology." *Circulation Research* 95, no. 4 (2004): 343–53.

U.S. Congress. *Fetus Farming Prohibition Act of* 2006. S. 3504. 109th Cong., 2nd Sess. *U.S. Statutes at Large* 120, no. 570 (2006).

USA TODAY. "Our View on Medical Research: Stem Cell Reversal Fixes Bush's Flawed Compromise." March 10, 2009, A10.

Van de Velde, Hilde, Greet Cauffman, Herman Tournave, Paul Devroey, and Inge Liebaers. "The Four Blastomeres of a 4-Cell Stage Human Embryo are Able to Develop Individually into Blastocysts with Inner Cell Mass and Trophectoderm." *Human Reproduction* 23, no. 8 (2008): 1742–7.

Van der Elst, Josiane. "Oocyte Freezing: Here to Stay?" *Human Reproduction Update* 9, no. 5 (2003): 463–70.

Van Hoof, Dennis, Wilma Dormeyer, Stefan R. Braam, et al. "Identification of Cell Surface Proteins for Antibody-Based Selection of Human Embryonic Stem Cell-Derived Cardiomyocytes." *Journal of Proteome Research* 9, no. 3 (2010): 1610–8.

Van Inwagen, Peter. *Material Beings*. Ithaca, NY: Cornell University Press, 1990.

van Laake, Linda W., Robert Passier, Pieter A. Doevendans, and Christine L. Mummery. "Human Embryonic Stem Cell-Derived Cardiomyocytes and Cardiac Repair in Rodents." *Circulatory Research* 102, no. 9 (2008): 1008–10.

van Zyl, Bernard. *Adult Stem Cell Therapies Saved my Life*. Bloomington, IN: Rooftop Publishing, 2007.

Veltman, Caroline E., Osama I. I. Soliman, Marcel L. Geleijnse, et al. "Four-Year Follow-up of Treatment with Intramyocardial Skeletal Myoblasts Injection in Patients with Ischaemic Cardiomyopathy." *European Heart Journal* 29 (2008): 1386–96.

Verlinsky, Yury, Jacques Cohen, Santiago Munne, et al. "Over a Decade of Experience with Preimplantation Genetic Diagnosis: A Multicenter Report." *Fertility and Sterility* 82, no. 2 (2004): 292–4.

Vescell. http://archive.org/details/VesCellAdultStemCellTherapyforHeartDiseasebyVescell.

Vita, Marina, and Marie Henriksson. "The Myc Oncoprotein as a Therapeutic Target for Human Cancer." *Seminars in Cancer Biology* 16, no. 4 (2006): 318–30.

Voullaire, Lucille, H. Slater, R. Williamson, and L. Wilton. "Chromosome Analysis of Blastomeres from Human Embryos by Using Comparative Genomic Hybridization." *Human Genetics* 106 (2000): 210–7.

Vrana, Kent E., Jason D. Hipp, Ashley M. Goss, et al. "Nonhuman Primate Parthenogenetic Stem Cells." supplement, *Proceedings of the National Academy of Sciences* USA 100, no. S1 (2003): 11911–6.

Wakayama, Teruhiko, Viviane Tabar, Ivan Rodriguez, Anthony C. F. Perry, Lorenz Studer, and Peter Mombaerts. "Differentiation of Embryonic Stem Cell Lines Generated from Adult Somatic Cells by Nuclear Transfer." *Science* 292 (2001): 740–3.

Wakayama, Teruhiko. "Production of Cloned Mice and ES Cells from Adult Somatic Cells by Nuclear Transfer: How to Improve Cloning Efficiency?" *Journal of Reproduction and Development* 53, no. 1 (2007): 13–26.

Wake, Norio, Takahiro Arima, and Takao Matsuda. "Involvement of IGF2 and H19 Imprinting in Choriocarcinoma Development." *International Journal of Gynacology and Obstetrics* 60 (1998): S1–8.

Walker, Adrian J. "Altered Nuclear Transfer: A Philosophical Critique." *Communio* 31 (2005): 649–84.

Wang, Hanmin, Jijun Hao, and Charles C. Hong. "Cardiac Induction of Embryonic Stem Cells by a Small Molecule Inhibitor of Wnt/β-Catenin Signaling." *ACS Chemical Biology* 6, no. 2, (2010): 192–7.

Wang, Yigang, Dongsheng Zhang, Muhammad Ashraf, et al. "Combining Neuropeptide Y and Mesenchymal Stem Cells Reverses Remodeling after Myocardial Infarction." *American Journal of Physiology, Heart and Circulatory Physiology* 298, no. 1 (2010): H275–86.

Watkins, William M., Amy D. Yates, Pamela Greenwell, et al. "A Human Dispermic Chimaera First Suspected from Analyses of the Blood Group Gene-Specified Glycosyltransferases." *Journal of Immunogenetics* 8 (1981): 113–28.

Watson, Robert A., and Trevor M. Yeung. "What is the Potential of Oligodendrocyte Progenitor Cells to Successfully Treat Human Spinal Cord Injury?" *BMC Neurology* 11 (2011): 113.

Waxman, Stephen G. "Demyelination in Spinal Cord Injury." *Journal of the Neurological Sciences* 91 (1989): 1–14.

Wei, J. P., T. S. Zhang, S. Kawa. T. Aizawa, M. Ota, T. Akaike, K. Kato, I. Konishi, and T. Nikaido. "Human Amnion-Isolated Cells Normalize Blood Glucose in Streptozotocin-Induced Diabetic Mice." *Cell Transplantation* 12, no. 5 (2003): 545–52.

Weiss, Mark L., Satish Medicetty, Amber R. Bledsoe, et al. "Human Umbilical Cord Matrix Stem Cells: Preliminary Characterization and Effect of Transplantation in a Rodent Model of Parkinson's Disease." *Stem Cells* 24 (2006): 781–92.

Weiss, Rick. "A Crucial Human Cell Is Isolated, Multiplied; Embryonic Building Block's Therapeutic Potential Stirs Debate." *The Washington Post*, November 06, 1998, A01.

———. "Bush Unveils Bioethics Council; Human Cloning, Tests on Cloned Embryos Will Top Agenda of Panel's 1st Meeting." *Washington Post*, January 17, 2002, A21.

Wells, David N., Pavla M. Misica, and H. Robin Tervit. "Production of Cloned Calves Following Nuclear Transfer with Cultured Adult Mural Granulosa Cells." *Biology of Reproduction* 60, (1999): 996–1005.

Wennberg, Robert. *Life in the Balance: Exploring the Abortion Controversy.* Grand Rapids, MI: Wm B. Eerdmans, 1985.

Whelan III, Joseph G., and Nikos F. Vlahos. "The Ovarian Hyperstimulation Syndrome." *Fertility and Sterility* 73, no. 5 (2000): 883–96.

White, Robin E., Meghan Rao, John C. Gensel, Dana M. McTigue, Brian K. Kaspar, and Lyn B. Jakeman. "Transforming Growth Factor α Transforms Astrocytes to a Growth-Supportive Phenotype after Spinal Cord Injury." *Journal of Neuroscience* 31, no. 42 (2011): 15173–87.

Whittle, Wendy L., William Gibb, and John R. G. Challis. "The Characterization of Human Amnion Epithelial and Mesenchymal Cells: The Cellular Expression, Activity and Glucocorticoid Regulation of Prostaglandin Output." *Placenta* 21 (2000): 394–401.

Williams, Adam R., Barry Trachtenberg, Darcy L. Velazquez, et al. "Intramyocardial Stem Cell Injection in Patients with Ischemic Cardiomyopathy: Functional Recovery and Reverse Remodeling." *Circulation Research* 108, no. 7 (2011): 792–6.

Williamson, K., S. E. Stringer, and M. Y. Alexander. "Endothelial Progenitor Cells Enter the Aging Arena." *Frontiers in Vascular Physiology* 3 (2012): 30.

Wilmut, Ian, Angelika E. Schnieke, Jim McWhir, Alexander J. Kind, and Keith H. S. Campbell. "Viable Offspring Derived from Foetal and Adult Mammalian Cells." *Nature* 385 (1997): 810–3

Wilmut, Ian, Keith Campbell, and Colin Tudge. *The Second Creation; Dolly and the Age of Biological Cloning*. New York: Farrar, Straus and Giroux, 2000.

Wilson, Anne, and Andreas Trumpp. "Bone-Marrow Haematopoietic-Stem-Cell Niches." *Nature Reviews Immunology* 6 (2006): 93–106.

Wilson, Kitchener D., Shivkumar Venkatasubrahmanyam, Fangjun Jia, Ning Sun, Atul J. Butte, and Joseph C. Wu. "MicroRNA Profiling of Human-Induced Pluripotent Stem Cells." *Stem Cells and Development* 18, no. 5 (2009): 749–59.

Winitsky, Steve O., Thiru V. Gopal, Shahin Hassanzadeh, et al. "Adult Murine Skeletal Muscle Contains Cells That Can Differentiate into Beating Cardiomyocytes In Vitro." *PLoS Biology* 3, no. 4 (2005): e87. doi:10.1371/journal.pbio.0030087.

Wong, Connie C., Kevin E Loewke, Nancy L. Bossert, Barry Behr, Christopher J. De Jonge, Thomas M. Baer, Renee A. Reijo Pera. "Non-Invasive Imaging of Human Embryos Before Embryonic Genome Activation Predicts Development to the Blastocyst Stage." *Nature Biotechnology* 28, no. 10 (2010): 1115–21.

Woods, Gordon L., Kenneth L. White, Dirk K. Vanderwall, et al. "A Mule Cloned from Fetal Cells by Nuclear Transfer." *Science* 301 (2003): 1063.

Wright, Karina T., Wagih El Masri, Aheed Osman, Joy Chowdhury, and William E. B. Johnson. "Concise Review: Bone Marrow for the Treatment of Spinal Cord Injury: Mechanisms and Clinical Applications." *Stem Cells* 29 (2011): 169–78.

Xiang, Guosheng, Qing Yang, Bing Wang, et al. "Lentivirus-mediated Wnt11 Gene Transfer Enhances Cardiomyogenic Differentiation of Skeletal Muscle-derived Stem Cells." *Molecular Therapy* 19 (2011): 790–6.

Xie, Chang-Qing, Huarong Huang, Sheng Wei, et al. "A Comparison of Murine Smooth Muscle Cells Generated from Embryonic Versus Induced Pluripotent Stem Cells." *Stem Cells and Development* 18, no. 5 (2009): 741–8.

Xie, Xinxing, Aijun Sun, Wenqing Zhu, Zheyong Huang, Xinying Hu, Jianguo Jia, Yunzeng Zou, and Junbo Ge. "Transplantation of Mesenchymal Stem Cells Preconditioned with Hydrogen Sulfide Enhances Repair of Myocardial Infarction in Rats." *Tohoku Journal of Experimental Medicine* 226 (2012): 29–36.

Xu, Dan, Zaida Alipio, and Louis M. Fink. "Phenotypic Correction of Murine Hemophila A Using an iPS Cell-Based Therapy." *Proceedings of the National Academy of Sciences USA* 106, no. 3 (2009): 808–13.

Xu, Yi, Masaaki Kitada, Masahiro Yamaguchi, Mari Dezawa, and Chizuka Ide. "Increase in bFGF-Responsive Neural Progenitor Population Following Contusion Injury of The Adult Rodent Spinal Cord." *Neuroscience Letters* 397 (2006): 174–9.

Yan, Jun, Leyan Xu, Annie M. Welsh, et al. "Extensive Neuronal Differentiation of Human Neural Stem Cell Grafts in Adult Rat Spinal Cord." *PLoS Medicine* 4, no. 2 (2007): e39. doi:10.1371/journal.pmed.0040039.

Yang, Junjie, Masaaki Ii, Naosuke Kamei, et al. "CD34+ Cells Represent Highly Functional Endothelial Progenitor Cells in Murine Bone Marrow." *PLoS One* 6, no. 5 (2011): e20219. doi:10.1371/journal.pone.0020219.

Yang, K. L., C. Y. Chang, S. Lin, M. H. Shyr, and P. Y. Lin. "Unrelated Haematopoietic Stem Cell Transplantation in Taiwan and Beyond." supplement, *Hong Kong Medical Journal* 15, no. S3 (2009): 48–51.

Yang, Xiangzhong, Sadie L Smith, X. Cindy Tian, Harris A. Lewin, Jean-Paul Renard, and Teruhiko Wakayama. "Nuclear Reprogramming of Cloned Embryos and Its Implications for Therapeutic Cloning." *Nature Genetics* 39 (2007): 295–302.

Yang, Yang, Yan Zeng, Zhuo Lv, et al. "Abnormal Development at Early Postimplantation Stage in Mouse Embryos After Preimplantation Genetic Diagnosis." *Anatomical Record* 295 (2012): 1128–33.

Yang, Zhijian, Fumin Zhang, Wenzhu Ma, et al. "A Novel Approach to Transplanting Bone Marrow Stem Cells to Repair Human Myocardial Infarction: Delivery Via a Noninfarct-Relative Artery." *Cardiovascular Therapies* 28, no. 6 (2010): 380–5.

Yao, Shuyuan, Shuibing Chen, Julie Clark, et al. "Long-Term Self Renewal and Directed Differentiation of Human Embryonic Stem Cells in Chemically Defined Conditions." *Proceedings of the National Academy of Sciences* USA 103 (2006): 6907–12.

Yoon, Jihyun, Byoung Goo Min, Young-Hoon Kim, Wan Joo Shim, Young Moo Ro, and Do-Sun Lim. "Differentiation, Engraftment and Functional Effects of Pre-Treated Mesenchymal Stem Cells in a Rat Myocardial Infarct Model." *Acta Cardiologica* 60, no. 3 (2005): 277–84.

Yoon, Young-sup, Andrea Wecker, Lindsay Heyd, et al. "Clonally Expanded Novel Multipotent Stem Cells from Human Bone Marrow Regenerate Myocardium after Myocardial Infarction." *Journal of Clinical Investigation* 115, no. 2 (2005): 326–38.

Young, Lorraine E., Kevin D. Sinclair, and Ian Wilmut. "Large-Offspring Syndrome in Cattle and Sheep." *Reviews of Reproduction* 3 (1998): 155–63.

Young, Pampee P., Douglas E. Vaughan, and Antonis K. Hatzopoulos. "Biological Properties of Endothelial Progenitor Cells (EPCs) and their Potential for Cell Therapy." *Progress in Cardiovascular Dis*eases 49, no. 6 (2007): 421–9.

Young, Randell G., David L. Butler, Wade Weber, Arnold I. Caplan, Stephen L. Gordon, and David J. Fink. "Use of Mesenchymal Stem Cells in a Collagen Matrix for Achilles Tendon Repair." *Journal of Orthopaedic Research* 16, no 4 (1998): 406–13.

Young, Robert. *Young's Literal Translation of the Bible*. Laguna Hills, CA: OSNOVA, 2010. Kindle edition.

Yu, Junying, and James A. Thomson. "Pluripotent Stem Cell Lines." *Genes and Development* 22 (2008): 1987–97.

Yu, Junying, Maxim A. Vodyanik, Kim Smuga-Otto, et al. "Induced Pluripotent Stem Cell Lines Derived from Human Somatic Cells." *Science* 318 (2007): 1917–20.

Yu, Neng, Margot S. Kruskall, Juan J. Yuni, et al. "Disputed Maternity Leading to Identification of Tetragametic Chimerism." *New England Journal of Medicine* 346 (2002): 1545–52.

Yu, Yang, Jindao Wu, Yong Fan, et al. "Evaluation of Blastomere Biopsy Using a Mouse Model Indicates the Potential High Risk of Neurodegenerative Disorders in the Offspring." *Molecular & Cell Proteomics* 7 (2009): 1490–500.

Zacharias, Ravi, and Laure Fournier. *The Merchant and the Thief: A Folktale from India*. Grand Rapids, MI: Zonderkidz, 2012.

Zakhartchenko, Valeri, Ramiro Alberio, Miodrag Stojkovic, et al. "Adult Cloning in Cattle: Potential of Nuclei from a Permanent Cell Line and from Primary Cultures." *Molecular Reproductive Development* 54 (1999): 264–7.

Zeher, Margit, Gabor Papp, and Peter Szodoray. "Autologous Haemopoietic Stem Cell Transplantation for Autoimmune Diseases." *Expert Opinion on Biological Therapy* 11, no. 9 (2011): 1193–201.

Zhang, Jianhua, Gisela F. Wilson, Andrew G. Soerens, et al. "Functional Cardiomyocytes Derived from Human Induced Pluripotent Stem Cells." *Circulation Research* 104 (2009): e30–e41.

Zhang, Liang, Hong-Tian Zhang, Sun-Quan Hong, Xu Ma, Xiao-Dan Jiang and Ru-Xiang Xu. "Cografted Wharton's Jelly Cells-Derived Neurospheres and BDNF Promote Functional Recovery after Rat Spinal Cord Transection." *Neurochemical Research* 34 (2009): 2030–9.

Zhang, Qiangzhe, Junjie Jiang, Pengcheng Han, et al. "Direct Differentiation of Atrial and Ventricular Myocytes from Human Embryonic Stem Cells by Alternating Retinoid Signals." *Cell Research* 21, no. 4 (2011): 579–87.

Zhang, Xin, Petra Stojkovic, Stefan Przyborski et al. "Derivation of Human Embryonic Stem Cells from Developing and Arrested Embryos." *Stem Cells* 24 (2006): 2669–76.

Zhang, Ying Ming, Criss Hartzell, Michael Narlow, and Samuel C. Dudley Jr. "Stem-Cell Derived Cardiomyocytes Demonstrate Arrhythmic Potential." *Circulation* 106 (2002): 1294–9.

Zhang, Zheng, Hu Lin, Ming Shi, et al. "Human Umbilical Cord Mesenchymal Stem Cells Improve Liver Function and Ascites in Decompensated Liver Cirrhosis Patients." supplement, *Journal of Gastroenterology and Hepatology* 27, no. S2 (2012): 112–20.

Zhang, Zhijie, Preston Burnley, Brandon Coder, and Dong-Ming Su. "Insights on *Foxn1* Biological Significance and Usages of the 'Nude' Mouse in Studies of T-Lymphopoiesis." *International Journal of Biology* 8 (2012): 1156–67.

Zhao, Tongbiao, Zhen-Ning Zhang, Zhili Rong, and Yang Xu. "Immunogenicity of Induced Pluripotent Stem Cells." *Nature* 474 (2011): 212–5.

Zhou, Hongyan, Shili Wu, Jin Young Joo, et al. "Generation of Induced Pluripotent Stem Cells Using Recombinant Proteins." *Cell Stem Cell* 4, no. 5 (2009): 381–4.

Zhou, Qi, Jean-Paul Renard, Gaëlle Le Friec, et al. "Generation of Fertile Cloned Rats by Regulating Oocyte Activation." *Science* 302 (2003): 1179.

Zhou, Qiao, Juliana Brown, Andrew Kanarek, Jayaraj Rajagopal, and Douglas A. Melton. "In Vivo Reprogramming of Adult Pancreatic Exocrine Cells into β-Cells." *Nature* 455, (2008): 627–33.

Zhou, Hongyan, and Sheng Ding. "Evolution of Induced Pluripotent Stem Cell Technology." *Current Opinion in Hematology* 17, no. 4 (2010): 276–80.

Zhou, Ting, Christina Benda, Sarah Duzinger, et al. "Generation of Induced Pluripotent Stem Cells from Urine." *Journal of the American Society of Nephrology* 22, no. 7 (2011): 1221–8.

Ziegner, Ulrike H. M., Hans D. Ochs, Carolyn Schanen, et al. "Unrelated Umbilical Cord Stem Cell Transplantation for X-Linked Immunodeficiencies." *Journal of Pediatrics* 138, no. 4 (2001): 570–3.

Zvaifler, Nathan J., Lilla Marinova-Mutafchieva, Gill Adams, et al. "Mesenchymal Precursor Cells in the Blood of Normal Individuals." *Arthritis Research* 2 (2000): 477–88.

Index